COMPREHENSIVE SEWER COLLECTION SYSTEMS ANALYSIS HANDBOOK FOR ENGINEERS AND PLANNERS

COMPREHENSIVE SEWER COLLECTION SYSTEMS ANALYSIS HANDBOOK FOR ENGINEERS AND PLANNERS

Nicklow/Boulos/Muleta

MWH Soft, Inc.

MWH Soft, Inc.

COMPREHENSIVE SEWER COLLECTION SYSTEMS ANALYSIS HANDBOOK FOR ENGINEERS AND PLANNERS

First Edition

Authors

John W. Nicklow, Ph.D
Associate Professor
Southern Illinois University at Carbondale

Paul F. Boulos, Ph.D
President and Chief Operating Officer
MWH Soft, Inc.

Misgana K. Muleta, Ph.D
Senior Engineer
MWH Soft, Inc.

MWH SOFT, INC.
Pasadena, California USA

COMPREHENSIVE SEWER COLLECTION SYSTEMS ANALYSIS HANDBOOK FOR ENGINEERS AND PLANNERS

FIRST EDITION

ISBN 0-9745689-1-0

MWH Soft, Inc.
300 North Lake Avenue, Suite 1200
Pasadena, California 91101
USA

Phone: +1-626-568-6868
Fax: +1-626-568-6870
Email: mwhsoft@mwhsoft.com
Internet: www.mwhsoft.com

To Sitt Marie-Rose

"Countless Children Lives Have Been Touched By Her Intercession – A Humanitarian Legend And An Inspiration To Us All"

TABLE OF CONTENTS

CHAPTER 4. UNSTEADY FLOW HYDRAULICS 4-1

CHAPTER 5. WATER QUALITY ANALYSIS 5-1

PREFACE

This book is concerned with the simulation and analysis of sanitary, storm and combined sewer collection systems. A sanitary sewer transports waterborne wastes containing some minor quantities of inadvertent storm, surface, and groundwater flows from residences, commercial buildings, industrial plants, and institutions. A storm sewer transports storm runoff along with street waste and wash water or drainage. A combined sewer carries a mixture of domestic and industrial wastewater along with storm runoff. Typically, combined sewer systems are found in older and more densely built urban cities, while separated sanitary and stormwater systems are found in the more recently developed suburban areas.

Billions of dollars continue to be spent at all levels of government in the United States toward the rehabilitation or replacement of the nation's 19,800 sewer collection systems. Leaking sanitary sewers can cause exfiltration of raw wastewater and industrial discharge contaminating soil and groundwater. Infiltration of groundwater through defective joints, broken pipes, improper connections, or manhole walls can lead to excess cost at the treatment plant or contribute to pipeline failures. Storm conditions can cause the capacity of the combined sewer to be exceeded resulting in a surcharged sewer system with potential backups, flood damage and health threats. Combined sewer overflows often result in significant discharges of organic material, nutrients, sediment, microorganisms, oil, metals and other potentially toxic substances. Stormwater can also find its way into a sanitary sewer system from illegal connections, cracks, and leaks through maintenance hole covers leading to sanitary sewer overflows. The adoption by Congress of the Clean Water Act gave the U.S. Environmental Protection Agency the authority to implement pollution control programs and set wastewater standards for the industry. Comprehensive modeling and analysis of these sewer conveyance systems becomes necessary for developing sound, cost-effective remedial alternatives for enhancing system integrity and performance, restoring and maintaining needed capacity, avoiding backups and overflows, meeting environmental regulations, and improving public health and safety.

Most existing textbooks provide general information focusing on the fundamental theory of open channel hydraulics with little or no emphasis on practical modeling applications. In addition, engineering courses may cover the general principles governing the hydraulics and design of sewer collection systems, but usually do not explicitly consider dry and wet weather load estimation; infiltration/inflow contribution; rainfall-runoff modeling; water quality analysis; and overall network modeling applications. Consequently, many consultants and wastewater utility engineers are not familiar with sewer collection systems modeling to the extent needed to support a sound capital improvement plan. The authors feel that a need exists for a book that brings together both the theoretical and practical information required for the effective modeling of hydraulic and water quality conditions in sewer systems. This book was written in an attempt to help satisfy the needs of both the practicing engineer and the engineering student. Throughout, focus is on actual implementation of concepts and various applications. The book provides a comprehensive overview of sewer collection systems modeling from steady uniform flow to varied flow to more sophisticated unsteady flow analyses, encompassing both hydraulic and flow quality network modeling applications. The aim is to develop in students and practicing engineers deep understanding of sewer system management activities and their links to performance and safety. We feel this objective is best achieved by capturing the essence of sewer collection systems modeling – the fundamental principles upon which it is based, the practical problems in conceptualizing real-world situations in model form, and the use of models to guide decision making.

This book was written in the spirit of promoting a thorough understanding of the tools necessary to model and analyze sewer collection systems. It begins with a fundamental review of collection systems and their components, followed by detailed guidelines for estimating sewer loads. Methods are presented for both steady and unsteady hydraulic analysis of sewer flows and for the analysis of flow quality. Discussion of associated stormwater detention and retention facilities, review of sewer system regulatory requirements, and discussion of urban watershed management techniques are also presented. These various topics are described in a manner meant to be easily understood and that focuses on the application of principles. The book contains over 100 worked examples that are helpful in understanding basic principles and their practical applications. As such, it serves as a practical guide and a source of useful information for use by engineers, municipalities, regulatory agencies, and others with responsibility for planning, designing, constructing, financing, regulating, operating, and maintaining sanitary, storm and combined sewer systems. The book is arranged into eight chapters and is written using both SI and U.S. customary units. Its objective is to present the basic principles of flow in open channels and sewer collection engineering and the application of these principles to solving practical, real-life problems. The book can be used to teach sewer system modeling at both the undergraduate and graduate levels, as well as an excellent reference for any wastewater utility professional. It is an invaluable reference to accumulate knowledge and find comprehensive sewer network modeling and design concepts and procedures, worked-out example problems, and up-to-date references for further reading. Combining theory with hands-on modeling applications, the book provides a complete and enriching educational experience, even to readers new to the wastewater industry.

We have drawn material from many sources and wish to thank them all.

ACKNOWLEDGMENTS

This book is a product of a learning journey and our debts are great. During the past several years, we have worked with some of the world's finest utilities and exceedingly bright university professors, engineers, operators and regulators, and we have learned a great deal from them. This book is an outgrowth of this extensive experience in sewer infrastructure engineering and could not have been completed without the assistance and sound advice of many extraordinarily insightful and selfless friends, colleagues, students and associates. Their wisdom, teaching and body of knowledge are embedded in the heart of our work. To all of them, we extend our sincere thanks. The strengths of this book are theirs, its weaknesses our own.

We are greatly indebted to Richard Wankmuller, Robert Uhler, Donald Smith, Murli Tolaney, Rudy Tekippe, Daniel McConville, Donal Bassett, Vic Gulas, Kevin Kelly, Donald Evenson, Jeff Kishel, Ed Wetzel, and Ian Noble of MWH for providing a wonderful working environment and for being a source of invaluable support and endless encouragement. To say the least, their energy and enthusiasm were contagious and have fashioned our best efforts into the book you are now holding. They were an author's dream.

We are also particularly grateful to the team of talented professionals at MWH Soft who participated directly or indirectly, but all meaningfully, in the preparation of this book. Our debt to them as teachers and friends is truly incalculable.

You don't write a book without some special people very close to you supporting you and believing that you can. Our deepest gratitude is to our families, especially our wives, Stacy, Katya, and Obse, for their unceasing love, faith and emotional support and for the many hours that should have been spent with them but were required in the preparation of this book. They have been and continue to be our true partners and our source of inspiration – we owe them so much.

Preparing this book has been a labor of passion and love for us, and it is our fervent hope that it will inspire and serve our profession well in the years to come.

We dedicate this book to world peace, humanity and human welfare. And to the children in our lives who are our best hopes and our fondest dreams.

John W. Nicklow
Paul F. Boulos
Misgana K. Muleta

CHAPTER ONE
Introduction

1.1 OVERVIEW OF SEWER COLLECTION SYSTEMS

1.1.1 The Need

Population growth and urban development create distinct problems with respect to stormwater and wastewater. First, construction of homes, commercial and industrial developments, parking lots, and streets reduce the natural infiltration capacity of a drainage area and alter surface flow patterns. These factors, in turn, increase the volume and velocity of stormwater runoff and generally produce larger peak discharges. Since urban land surfaces are subject to the accumulation of pollutants, the quality of runoff is also impacted. These nonpoint source pollutants are transported and can contribute significantly to reduced quality of receiving waters. Secondly, increased water demand by those residential, commercial, and industrial establishments translates into an increased volume of generated wastewater. The solution to these problems, at least in part, lies in the design and construction of adequate sewer systems that collect and convey stormwater runoff and wastewater flows for treatment, in the case of wastewater, and for subsequent disposal. In many cases, sewers must also convey quantities of groundwater that leak into the collection system.

This book is intended to give the reader a basic understanding of the tools necessary to analyze sewer systems. It begins with a fundamental review of collection systems and their components, followed by detailed guidelines for estimating sewer loads. Methods are presented for both steady and unsteady hydraulic analysis of sewer flows and for the simulation of flow quality. Discussion of associated stormwater detention and retention facilities, review of sewer collection system regulations, and discussion of urban watershed management techniques are also presented. These various topics are described in a manner meant to be easily understood and that focuses on the application of principles (e.g., over 100 solved problems). As such, it serves as a practical guide for engineers and planners involved in the analysis, design and operation of sewer collection systems.

1.1.2 Types of Sewer Systems

There are three types of sewer collection systems: storm sewers, sanitary sewers and combined sewers. Storm sewers (Figure 1-1) are designed to collect and transport stormwater runoff, also referred to as wet weather flow, which is produced when the capacity of the land to retain precipitation is exceeded. Surface runoff is then collected and enters sewers through a carefully-designed network of roadway gutters and street inlets (e.g., grate, curb opening and combination inlets). Operation of the collection system thus depends on wet weather conditions, such as storm intensity and duration, and characteristics of the drainage area, including slope, land use, soil type, and degree of imperviousness. These systems are almost always designed to flow partially full or under just-full conditions. Thus, runoff travels through the sewer as open channel flow, in which gravity forces dominate movement. Disposal of flow occurs within existing watercourses with little to no treatment. Storm sewers often include impoundments such as detention or retention facilities that temporarily store flood waters so that peak post-development runoff rates are reduced to levels less than or equal to peak pre-development rates. These facilities also tend to improve water quality by allowing pollutants to settle.

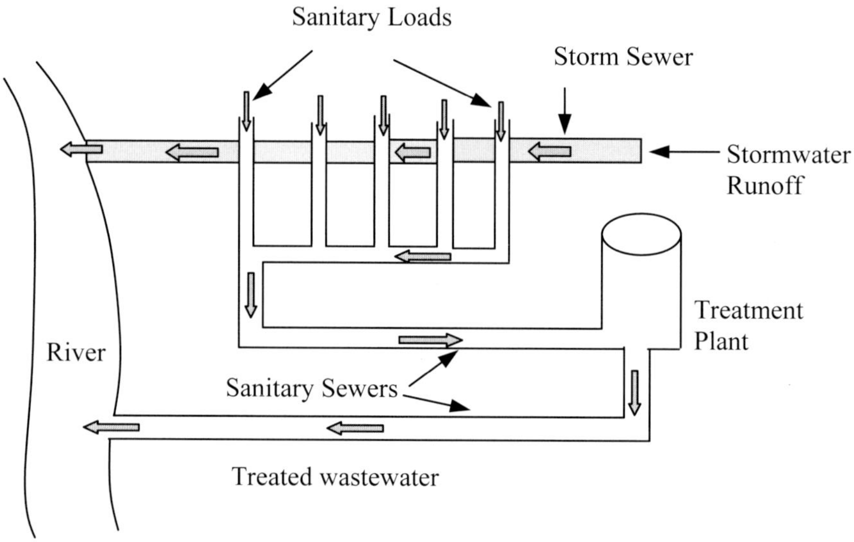

Figure 1-1: Separate sanitary and stormwater collection systems

Sanitary sewers (Figure 1-1) are designed to collect and convey wastewater, or dry weather flow, from industrial, commercial, institutional, or residential areas. The discharge typically consists of highly concentrated organic matter and requires treatment prior to disposal and subsequent return to the natural environment. Because wastewater volumes are typically less than comparable stormwater flows, sanitary systems are generally smaller in size. Similar to storm sewers, however, they are most often designed to flow under gravity conditions. In fact, due to the uncertainty associated with wastewater loads, sanitary sewers are often designed to flow between one-half and three-fourths full. These gravity sewers allow for the transfer of oxygen across the free surface, and aerobic conditions may exist. Alternatively, smaller diameter sanitary sewer lines that are designed to operate under pressure are called pressure or force mains. Anaerobic conditions generally dominate in these systems since re-aeration is significantly limited. Force mains are required where variations in topography or excavation depths preclude the use of gravity sewer mains.

In combined sewer systems (Figure 1-2), wastewater and stormwater runoff are collected and transported together. Since combined sewer systems operate during both wet weather and dry weather conditions, chemical and biological makeup of flow is highly variable. Furthermore, during severe storms, it is not feasible to treat the entire sewer load. Excess diluted flows are passed to a watercourse through regulators, or overflow structures (e.g., side weir or siphon), thus introducing higher pollutant levels in receiving waters. Although few such sewer systems are now being constructed due to water quality impacts and higher operational costs for treatment, many older cities and communities maintain combined sewer collection systems that were constructed years ago.

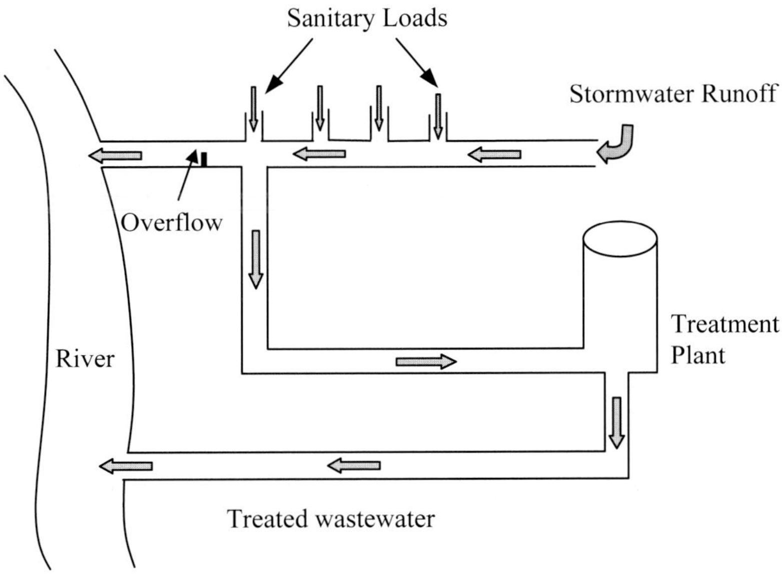

Figure 1-2: Combined sewer collection systems

In the US, there remain approximately 1,100 combined sewer systems and 10,000 to 15,000 combined sewer overflow discharge points in 772 cities serving about 40 million people.

1.1.3 Components of Sewer Systems

The primary component of all collection systems is the sewer conduit itself. Conduit is typically circular in cross section, although a variety of shapes can, and have been, used. In addition, systems typically contain a number of manholes, as shown in Figure 1-3. Manholes serve as access points to allow for maintenance and cleaning of the system and are generally installed at all changes in slope, pipe size, and horizontal and vertical alignments, as well as at all junctions and at the end of a sewer main. They are typically 4 ft (1.2 m) in diameter within the chamber, with a 2 ft (0.6 m) diameter frame and cover, and are spaced between 300 and 500 ft (90 and 150 m) apart. The manhole invert conforms, to the extent possible, to the shape and slope of the entering and exiting pipes. Drop manholes are installed when the sewer entering a manhole lies 2 ft (0.6 m) or more above the invert. In some cases, junctions and manholes have the capacity to temporarily store flows (i.e., storage junction); whereas a point junction has no capacity for storage.

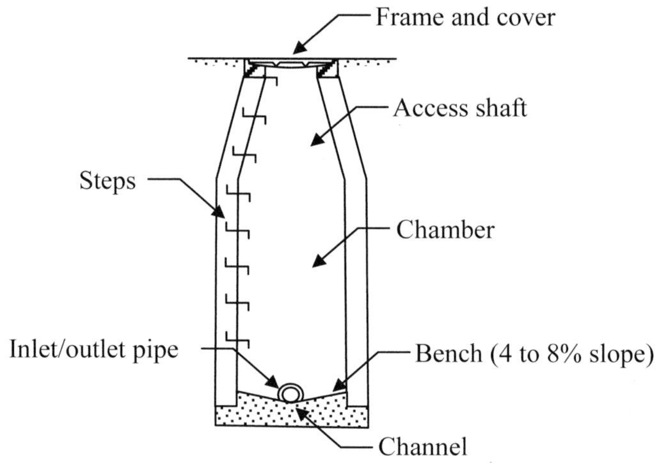

Figure 1-3: Typical manhole (adapted from Brown et al., 1996)

Depending on site conditions, other possible system components include pump stations and inverted siphons. Pump stations, also called lift stations, are frequently used in flat terrain to raise flow to a higher elevation so that gravity conditions can continue at reasonable slopes. Inverted siphons, or depressed sewers, are used for crossing streams or low lying regions. Within the depressed region of the sewer profile, the pipe remains full and under pressure, even without flow. In addition, for sanitary systems, cleanouts may be installed near the upstream ends of sewers to allow for cleaning or flushing. Storm water systems, alternatively, may include detention or retention basins (see Chapter 6) and street inlets.

1.2 FUNDAMENTAL ASPECTS OF SEWER ANALYSIS

1.2.1 Fluid Properties

The pertinent physical properties used to characterize sewer discharge include mass density, specific weight, and viscosity. These become important quantities within the governing laws that define fluid motion. Mass density, ρ, is the mass of fluid per unit volume. For practical purposes, storm and sanitary sewer flows have densities equivalent to that of water, as listed in Table 1-1. Though density increases slightly with decreasing temperature, a typical value used in analysis is 1.94 slugs/ft^3 (1,000 kg/m^3), where 1 slug is equivalent to 32.2 lbm.

Specific weight, γ, is the weight of fluid per unit volume. For water at normal temperatures, γ is approximately 62.4 lb/ft^3 (9,810 N/m^3). Through Newton's Second Law of Motion, specific weight is related to density by

$$\gamma = \rho g \tag{1-1}$$

where g is acceleration due to gravity, equal to 32.2 ft/s^2 (9.81 m/s^2). It is common to express the relative specific weight or density of a fluid to that of water at 39.4°F (4°C), a ratio that is known as specific gravity, S. For example, an estimate of S for wastewater is 1.0025, indicating that it is 0.25 percent heavier than water, a difference that is negligible for most applications.

Dynamic viscosity of a fluid, μ, represents a fluid's ability to resist deformation caused by shear forces. The higher the viscosity, the greater shear stresses that will develop within the fluid and the greater the friction losses that will occur along the sewer pipe wall. In most fluid flow applications, viscosity appears with density as μ/ρ, which is referred to as kinematic viscosity, ν. For liquids, viscosity decreases with increasing temperature.

1.2.2 Laws of Conservation

Although characteristics of the fluid are important, they cannot alone describe a physical problem. As with any natural or engineered system, flow through sewers occurs in accordance with a set of governing laws that contribute significantly to the nature and extent of the system's response. These laws, discussed in Chapters 3 and 4, imply conservation of mass, energy and momentum throughout the system. Although different mathematical forms and expressions are possible for particular applications and assumptions, these conservation laws apply in a general sense to every case.

Table 1-1: Physical properties of water

Temperature	Density (ρ)	Specific weight (γ)	Dynamic viscosity (μ)	Kinematic viscosity (ν)
(°F)	(slugs/ft^3)	(lb/ft^3)	(lb-s/ft^2)	(ft^2/s)
40	1.94	62.43	3.23×10^{-5}	1.66×10^{-5}
50	1.94	62.40	2.73×10^{-5}	1.41×10^{-5}
60	1.94	62.37	2.36×10^{-5}	1.22×10^{-5}
70	1.94	62.30	2.05×10^{-5}	1.06×10^{-5}
80	1.93	62.22	1.80×10^{-5}	9.30×10^{-6}
100	1.93	62.00	1.42×10^{-5}	7.39×10^{-6}
120	1.92	61.72	1.17×10^{-5}	6.09×10^{-6}
140	1.91	61.38	9.81×10^{-6}	5.14×10^{-6}
(°C)	(kg/m^3)	(N/m^3)	(N-s/m^2)	(m^2/s)
0	1000	9810	1.79×10^{-3}	1.79×10^{-6}
10	1000	9810	1.31×10^{-3}	1.31×10^{-6}
20	998	9790	1.00×10^{-3}	1.00×10^{-6}
30	996	9771	7.97×10^{-4}	8.00×10^{-7}
40	992	9732	6.53×10^{-4}	6.58×10^{-7}
50	988	9693	5.47×10^{-4}	5.53×10^{-7}
60	983	9643	4.66×10^{-4}	4.74×10^{-7}

Source: Adapted from Crowe et al. (2001)

1.2.3 Hydrologic Processes

The hydrologic cycle wholly describes the movement and distribution of water through the atmosphere (i.e., precipitation), over and through land surfaces (i.e., direct runoff and infiltration), to the ocean (i.e., discharge), and back to the atmosphere (i.e., evaporation). The cycle is driven primarily by solar energy and proceeds endlessly. With respect to stormwater collection systems analysis, focus is on direct or surface runoff, or that portion of precipitation that is not stored or infiltrated to the subsurface. The engineer is especially concerned with simulating the transformation of precipitation to discharge hydrographs, which describe the variation of sewer loads over time at a particular location.

Various methods exist for estimating runoff (see Chapter 2) or wet weather loads. They include empirical and semi-empirical techniques, such as the rational method for smaller drainage areas and hydrograph analysis for larger drainage areas, or physically-based approaches. The latter modeling technique refers to computational models that are based on governing physical laws. While they are widely adaptable and more powerful than empirical techniques, physically-based models are generally more complex and require far more data for simulation of hydrologic processes.

1.2.4 Sewer System Networks

Depending on the type of system, sewer networks are generally comprised of a number of sewer laterals, sub-mains, mains, and intercepting lines, all arranged in a dendritic fashion (i.e., converging to a downstream location). Laterals serve as upstream units into which no other common pipe discharges. They are generally 4 in (10 cm) or larger and, for sanitary systems, connect to a house, commercial development, or industrial site. A submain, if needed, receives discharge from a number of laterals, and a sewer main collects flow from one or more submains. Sewer mains are typically 8 in (20 cm) or larger in diameter and generally follow street or utility easements. For sanitary systems, sewer mains are designed to convey peak flows of three to four times the daily-expected average flow, and a final interceptor sewer collects flow and conveys it to the point of treatment.

The principal materials used for sewer pipe are listed in Table 1-2. During selection of pipe for a given application, consideration should be given to flow composition. Of particular concern are sulfates, present in wastewater and found in most water supplies. If anaerobic conditions persist, bacterial reduction of sulfate generates hydrogen sulfide (H_2S) gas. The gas carries a strong, rotten-egg odor and is toxic in even low concentrations. When further oxidized by bacteria on exposed pipe walls, the gas produces sulfuric acid (H_2SO_4), which is highly corrosive to uncoated pipes constructed of cement-based and metal materials. Other considerations in selecting pipe materials should include soil characteristics, external loading (e.g., backfill, impact and live loads), installation requirements, abrasion conditions, and cost.

Criteria for sewer systems construction can vary by regional locale and municipality. Generally speaking, however, pipes that comprise a sewer system should maintain a self-cleansing velocity greater than 2 fps (0.6 m/s). Doing so is likely to prevent deposition of suspended material and subsequent loss of capacity. It also inhibits early decomposition of settled wastes and, thus, increased generation of hydrogen sulfide, for sanitary systems.

Table 1-2: Commonly used pipe materials

Pipe material	Description
Asbestos-cement	Light weight pipe for gravity or pressurized systems carrying non-septic and non-corrosive flow through non-corrosive soil. Available in diameters up to 42 in (1,050 mm), but its use has faded due to health hazards associated with its fabrication.
Concrete	Available in either reinforced or unreinforced sections and is widely used for larger diameter gravity or pressure mains. Reinforced pipe is available in diameters ranging from 12 in (300 mm) to 144 in (3,600 mm), while unreinforced sections are available in diameters of 4 in (100 mm) to 36 in (900 mm). Should not be selected for use with septic or corrosive flows or soils unless special linings are used.
Cast iron	Well suited for pressurized systems or for high velocities and large external loads. Special lining or coating is required for corrosive flows and soils. Available in diameters up to 48 in (1,200 mm).
Polyvinyl chloride (PVC), acrylonitrile-butadienestyrene (ABS), or polyethylene (PE)	Thermoplastics that are light weight and generally resistant to corrosive soils and wastewater constituents. Available primarily in smaller diameters up to 15 in (375 mm). Cannot withstand large external loads or deflections.
Vitrified clay	Resistant to acids, alkalines, hydrogen sulfide, and abrasion. Available in diameters of 4 in (100 mm) to 48 in (1,200 mm) and in standard or extra-strength classifications, but is brittle.
Corrugated metal	Storm sewer pipe that provides a long service life in a variety of soil and water conditions. Available in circular or arched cross sections and in diameters ranging from 6 in (150 mm) to 144 in (3,600 mm). Metallic, polymeric, or bituminous coatings are often applied, depending on soil and water chemistry.
Ductile iron	Long lasting, high strength pipe for gravity or pressurized systems. Susceptible to corrosion. Available in diameters ranging from 4 in (100 mm) to 64 in (1,600 mm).

Source: Adapted from Lindeburg (1992), Metcalf and Eddy (1981), AISI (1990), and others

Velocities greater than approximately 10 fps (3 m/s) should be avoided unless provisions against impact forces and abrasion are made. Bury or cover depths should be sufficient to allow for gravity flow at all entry points and to prevent freezing of pipes. In most cases, and depending on regional climate, between 4 ft (1.2 m) and 8 (2.4 m) ft of cover may be sufficient. In addition, buoyancy effects should be considered and flotation of the pipe should be prevented where high ground water levels are present.

1.2.5 Modeling of Sewer Systems

Computer modeling of sewer collection systems is the most viable and efficient means of simulating the effects of hydrologic and hydraulic processes on sewer systems, enabling evaluation of system performance as well as testing system improvement and control strategies. Computer models can be effectively used for planning, analysis, design, optimization, and real-time operational control of sewer collection systems.

These models, which are designed to symbolically represent and describe behavior of flow through sewer systems, must perform three basic operations: (1) estimate sewer loadings; (2) evaluate system hydraulics; and (3) analyze water quality conditions. However, before an analysis can be initiated, adequate data must be collected.

1.2.5.1 Data Requirements

Modeling a collection system starts by acquiring data that describes existing site conditions, as well as future development information. Sources for data range from various local, state, and federal government agencies to pipe and pump manufacturers. Information required typically includes:

- Federal, state, and local regulatory standards and criteria;
- Watershed maps that identify topographic features and boundaries, as well as existing or expected drainage and land use patterns;
- Design and performance data for existing sewers and related facilities located upstream and downstream of the study location;
- Soil data identifying types, properties, and chemistry of soils;
- Intensity-duration-frequency and design storm data for storm sewers, and existing and planned residential, commercial, and industrial facilities and population data for sanitary systems;
- Data describing location of existing and proposed utilities;
- Right-of-way information and plan and profile data for roadways; and
- Information on various pipe materials used for sewer construction.

1.2.5.2 Load Estimation

The methods used to design and analyze storm and sanitary sewer systems are nearly identical. The major exception lies in the determination of design flows, or loading rates. For sanitary systems, flows are derived from measured or estimated wastewater loads from domestic, commercial, institutional, and industrial developments. In combined and storm sewer systems, an analysis of the hydrologic processes is necessary.

Peak loading rates should be used to evaluate whether a sewer system has sufficient capacity. Doing so will most likely prevent potentially damaging backflows or overflows. In addition, for combined and sanitary sewers, low flow conditions must be considered in order to limit solids deposition and odor generation.

1.2.5.3 Hydraulic Analysis

A quantitative evaluation of fluid movement and behavior throughout a collection system comprises the hydraulic analysis. It ultimately yields the spatial and temporal variation of velocity and flow depth at various locations in the sewer. Understanding this transient variation is important for assessing the performance and adequacy of the overall system.

Fundamentally, the hydraulic analysis is based on application of the laws of conservation. They govern flow through both sanitary and storm sewers, as well as combined systems, and apply to both pressure and gravity systems. Different expressed forms of these relationships are used, however, depending on whether conditions are assumed to be steady, in which velocity is constant over the time period considered (see Chapter 3), or unsteady (see Chapter 4), in which velocity changes overtime. In addition, relationships are required that define the impact on flow caused by pipe roughness and by bends, junctions, and other pipe appurtenances.

1.2.5.4 Water Quality Analysis

Water quality refers to the quantity of suspended and dissolved solids, dissolved gases, and microorganisms in a parcel of water. Analyses that are most useful in assessing the quality of sewer flows include an evaluation of flow age, source(s) tracing, simulating transport of conservative constituents, modeling of sediment movement and deposition, determination of biochemical oxygen demand (BOD), estimating the buildup and release of hydrogen sulfide, and prediction of rates of microbially induced corrosion. Each of these analyses is discussed in detail in Chapter 5.

Age of flow represents the time from when flow enters a system to when it exits. Since constituent concentrations depend on retention time in a system, age can be used as a general indication of overall quality and can be useful in identifying specific problem areas. For example, hydrogen sulfide generation is hastened in areas where wastewater is stale due to small slopes and velocities and, thus, increased flow age. Alternatively, source tracing refers to the determination of percentage of flows from any given source or particular location in the sewer. Tracing can be useful in evaluating spatial and temporal changes in loads or degree of mixing.

Conservative constituents such as sulfate, chloride, bromide, and trace metals are substances that remain in solution or suspension and are transported through sewer networks primarily through advection and, to a much lesser extent, dispersion. The former refers to the movement of a constituent with the fluid, moving in the direction of flow and with a velocity equivalent to that of the fluid, while the latter describes movement due to turbulence and is often neglected. Constituent transport and distribution are simulated using a mass balance of constituents coupled with an appropriate transport relationship. The simulation allows for an evaluation of mixing and the impact of constituents on receiving waters. Modeling the transport and deposition of suspended solids (i.e., sediments) is equally important, especially in sanitary sewer systems. Wastewater can carry substantial suspended solids, more than half of which are capable of settling. The analysis is critical in order to effectively prevent their accumulation and subsequent reduction in conveyance capacity of the sewer.

BOD is the amount of oxygen required to stabilize organic matter that is present in sewer flows. Discharges that contain significant amounts of biodegradable organic matter (e.g., wastewater) have high BOD levels and consume large amounts of oxygen. In turn, the dissolved oxygen in receiving aquatic systems will be adversely lowered by these discharges.

1.3 SOLVED PROBLEMS

Problem 1.1 Fluid properties

Consider a one-gallon can of paint recently purchased at a local hardware store. If the bucket is brim full and its contents weigh 15 lbf, compute the paint's mass, density, specific weight, and specific gravity.

Solution

The volume must first be converted to a consistent system of units.

$$1\ gallon \times \frac{1\ ft^3}{7.48\ gal.} = 0.134\ ft^3$$

From Newton's Second Law,

$$M = \frac{W}{g} = \frac{15}{32.2} = \underline{\underline{0.47\ slugs}}$$

Then density and specific weight can be evaluated as

$$\rho = \frac{M}{V} = \frac{0.47}{0.134} = \underline{\underline{3.5\ \frac{slugs}{ft^3}}}$$

$$\gamma = \frac{W}{V} = \frac{15}{0.134} = \underline{\underline{112\ \frac{lbs}{ft^3}}}$$

Finally, specific gravity is the dimensionless ratio of specific weight of the paint to that of water, or

$$S = \frac{\gamma_{paint}}{\gamma_{water}} = \frac{112}{62.4} = \underline{\underline{1.8}}$$

Problem 1.2 Fluid properties

If the specific weight of a liquid is 8,000 N/m^3, what are its density and specific gravity?

Solution

From Equation 1-1,

$$\rho = \frac{\gamma}{g} = \frac{8000}{9.81} = \underline{\underline{815\ \frac{kg}{m^3}}}$$

Then

$$S = \frac{\gamma_{fluid}}{\gamma_{water}} = \frac{8000}{9810} = \underline{\underline{0.82}}$$

Problem 1.3 Fluid properties

What is the percentage change in the specific weight of water when the temperature changes from 40°F to (a) 70°F and (b) 100°F?

Solution

At 40°F, γ is 62.43 lbs/ft^3. This value drops to 62.30 lbs/ft^3 at 70°F and 62.00 lbs/ft^3 at 100°F. In both cases, the percentage change is less than one percent (i.e., 0.2 percent for first case and 0.7 percent for the latter).

Comments: This problem illustrates the negligible effect of temperature change on specific weight, and thus density, at normal working temperatures that are encountered in the analysis of sewer collection systems.

Problem 1.4 Fluid properties

Consider that the specific gravity and kinematic viscosity of a certain liquid are 1.5 and 4×10^{-4} m^2/s, respectively. What is its dynamic viscosity?

Solution

Density of the liquid is computed as

$$\rho_{liquid} = S\rho_{water} = 1.5 \times 1000 = 1500 \ \frac{kg}{m^3}$$

Then,

$$\mu = \nu\rho = \left(4 \times 10^{-4}\right)\left(1500\right) = \underline{\underline{0.6 \ \frac{N \cdot s}{m^2}}}$$

Problem 1.5 Data sources

A number of government agencies collect and disseminate hydrologic data that are useful for individuals conducting storm water analyses. Using an Internet search engine or other means, identify some of the most commonly available sources for data.

Solution

Several of the most common sources include the United States Geologic Survey (USGS), National Weather Service (NWS), Environmental Protection Agency (EPA), United States Army Corps of Engineers (USACE), United States Bureau of Reclamation (USBR), and the Natural Resources Conservation Service (NRCS).

Problem 1.6 Steady vs. unsteady

Classify each of the following flows as steady or unsteady from the observer's viewpoint:

(a) Channel flow around bridge piers if the observer is
   (1) stationary, standing on the bridge;
   (2) in a drifting boat within the channel.

(b) Movement of storm water surge through a sewer if the observer is

(1) stationary;
(2) moving with the surge.

Solution

(a) Over short time periods, the stationary observer will see a steady flow, even though flow may be unsteady over longer time periods. The observer in the boat will see an unsteady flow as the boat passes under the bridge because velocity increases around the piers, even if the approach conditions are steady.

(b) A stationary observer will see unsteady flow as the surge passes. A steady flow will be observed if they move with the surge, assuming velocity is approximately constant over the distance considered.

REFERENCES CITED

American Iron and Steel Institute (AISI), *Modern Sewer Design*, 2nd ed., AISI, Washington, DC, 1990.

Brown, S.A., S.M. Stein, and J.C. Warner, *Urban Drainage Design Manual*, Hydraulic Engineering Circular No. 22, FHWA-SA-96-078, Federal Highway Administration, U.S. Department of Transportation, Washington, DC, 1996.

Crowe, C.T., D.E. Elger, and J.A. Roberson, *Engineering Fluid Mechanics*, 7th ed., Wiley, New York, NY, 2001.

Lindeburg, M.R., *Civil Engineering Reference Manual*, 6th ed., Professional Publications, Belmont, CA, 1992.

Metcalf and Eddy, Inc., *Collection and Pumping of Wastewater*, McGraw-Hill, New York, NY, 1981.

CHAPTER TWO
Load Estimation

2.1 INTRODUCTION

Sewer collection systems must be designed to safely convey anticipated peak discharges. During dry weather conditions, flows directly reflect water usage of the community and, therefore, can be expected to fluctuate significantly on an hourly, daily, or seasonal basis. For existing systems, the best data upon which to base sanitary sewer loads are actual records and flow measurements. In the absence of measured data, however, general guidelines exist for estimating flow rates. For storm and combined sewer systems, flows must be estimated based on a hydrologic analysis of excess precipitation and resulting runoff. Consider that for severe wet weather events, flow rates can increase by as much as a factor of 10^3 over the corresponding dry weather average.

2.2 DRY WEATHER LOAD ESTIMATION

Wastewater flows in sanitary sewers, and in combined sewers during dry weather periods, consist of three major components: (1) domestic wastewater; (2) industrial wastewater; and (3) infiltration and other inflows. Principal sources of domestic sanitary sewer loads are residential areas and commercial developments. Recreational and institutional facilities can contribute to large domestic loads as well. The average per capita domestic loading rate can be expected to vary between 50 and 265 gpd (190 and 1,000 Lpd) (ASCE, 1982). Peak daily flows, however, can range from two to four times greater than average daily flows. The wide range of flows reflects variations that can be caused by location, climate, community size, as well as related factors such as standard of living, water pricing, water quality, distribution system pressure, extent of meterage, and systems management (Fair et al., 1971). In the absence of measured data, Table 2-1 provides typical loading rates for various domestic sources. The design period throughout which the sewer capacity should be deemed adequate is between 25 and 50 years. Therefore, accurate population projections are essential in estimating future dry weather loads.

Industrial wastewater flows vary with the type and size of facility, as well as the degree of onsite treatment and reuse, if any. Typical values for facilities with little to no wet-process type industry range from 1,000 to 1,500 gpd/ac (9,350 to 14,020 Lpd/ha) for small developments and 1,500 to 3,000 gpd/ac (14,020 to 28,040 Lpd/ha) for medium-sized developments (Metcalf and Eddy,

1991). In cases where the specific nature of the industry is known, loads can be estimated on the basis of water usage data, such as that reported by Metcalf and Eddy (1991). For industries without water reuse or recycling facilities, it can be estimated that 85 to 95 percent of the water used in various plant operations will be returned as wastewater.

Table 2-1: Typical wastewater loads

Type of establishment	Lpd/person[a]	gpd/person[a]
Small dwellings and cottages with seasonal occupancy	190	50
Single-family dwellings	285	75
Multiple-family dwellings (apartments)	227	60
Rooming houses	150	40
Boarding houses	190	50
Additional kitchen wastes for nonresident boarders	38	10
Hotels without private baths	190	50
Hotels with private baths (2 persons per room)	227	60
Restaurants (toilet and kitchen wastes per patron)	26-38	7-10
Restaurants (kitchen wastes per meal served)	9-11	2.5-3
Additional for bars and cocktail lounges	8	2
Tourist camps or trailer parks (central bathhouse)	132	35
Tourist courts or mobile home parks (individual bath)	190	50
Resort camps (night and day) with limited plumbing	190	50
Luxury camps	380-570	100-150
Work or construction camps (semi-permanent)	190	50
Day camps (no meals served)	57	15
Day schools without cafeterias, gyms, or showers	57	15
Day schools with cafeterias, but no gyms or showers	75	20
Day schools with cafeterias, gyms, and showers	95	25
Boarding schools	285-380	75-100
Day workers at schools and offices (per shift)	57	15
Hospitals	570-945+	150-250+

[a] Unless otherwise noted in *type of establishment*

Table 2-1: Typical wastewater loads (*continued*)

Type of establishment	Lpd/person[a]	gpd/person[a]
Institutions other than hospitals	285-475	75-125
Factories (gal./pers./shift, apart from industrial waste)	57-132	15-35
Picnic parks (toilet wastes only)	19	5
Picnic parks with bathhouses, showers, & flush toilets	38	10
Swimming pools and bathhouses	38	10
Luxury residences and estates	380-570	100-150
Country clubs (per resident member)	380	100
Country clubs (per nonresident member present)	95	25
Motels (per bed space)	150	40
Motels with bath, toilet, and kitchen wastes	190	50
Drive-in theaters (per car space)	19	5
Movie theaters (per seat)	19	5
Airports (per passenger)	11-19	3-5
Self-service laundries (per wash)	190	50
Stores (per toilet room)	1500	400
Service stations (per vehicle served)	38	10

[a] Unless otherwise noted in *type of establishment*
Source: USPHS (1963)

Additional loads in sanitary sewers originate from infiltration and other steady inflows. Infiltration is that portion of water that enters the sewer from the ground through defective pipes or connections or through manhole walls. Quantities can range from 100 to 10,000 gpd/in-mile (9.3 to 930 Lpd/mm-km) of pipe (ASCE, 1982). The rate and volume of infiltration can be highly variable and depends on the length of sewers, number of connections, land area served, and soil and topographic conditions. In addition, infiltration quantities are dependent upon the quality of material and workmanship in construction of the system, maintenance practices, and the elevation of groundwater relative to the sewer. In particular, high groundwater tables capable of leaking into sewers can lead to large increases in wastewater flows, resulting in added costs for conveyance, treatment, and disposal. However, use of improved materials in modern sewer construction has significantly limited infiltration into newly-constructed sewers. Given its potential wide variability, the best estimation of infiltration is made by subtracting the normal 24-hour domestic and industrial loading rate from the measured 24-hour wastewater flow during dry weather periods.

Other dry weather inflows can include, but are not limited to, water from foundation drains, cooling-water discharges, and drains from springs or swamps. These loads represent steady inflows to the sewer that cannot be analyzed separately and, therefore, are often included in infiltration quantities.

2.2.1 Peak Flow Estimation Method

Sanitary sewers are normally sized on the basis of meeting projected design flows made up of peak wastewater flow and infiltration that are expected during the design period. The design period should normally outlast the bond issue or funding for the project. Average (or base) dry weather wastewater loads (Q_{base}) are transformed into peak loads (Q_{peaked}) using various load peaking methods. The two most common types of dry weather wastewater peak flow estimation methods are

$$Q_{Peaked} = KQ_{base}^{\rho} \tag{2-1}$$

and

$$Q_{peaked} = Q_{base}\left(\frac{a}{b+\left(\frac{P}{1000}\right)^{c}}+d\right) \tag{2-2}$$

where P represents the population; and a, b, c, d, K and ρ are peaking factor parameters. Equation 2-2 is the general form of the well known Babbitt and Harman coverage expressions. For the Babbitt equation, a= 5; b=0; c=0.2; and d=0, while for the Harman equation, a=14; b=4; c=0.5; and d=1 (Babbitt and Baumann, 1958). It should be noted that in the above equations, Q_{base} and P represent accumulated peakable flow and population, respectively.

2.3 *WET WEATHER LOAD ESTIMATION*

For storm sewer loading, the focus shifts to hydrologic analysis of excess precipitation and associated runoff. Common techniques for analysis include the rational method and unit hydrograph methods, as well as the use of more advanced hydrologic models. Wet weather loading for combined sewer systems is the same as for storm sewers; however, additional consideration must be given to wastewater flows and volumes. For example, long detention times during dry weather periods can lead to excess deposition and can cause septic conditions and odor problems. In addition, during severe storms, it is not cost effective to convey the entire mixture of wastewater and storm runoff to treatment works. It may be necessary, therefore, to reduce hydraulic loads by directing excess diluted flows to nearby streams through storm sewer overflows.

2.3.1 The Rational Method

For small drainage areas, peak runoff is commonly estimated by the rational method. This method is based on the principle that the maximum rate of runoff from a drainage basin occurs when all parts of the watershed contribute to flow and that rainfall is distributed uniformly over the catchment area. Since it neglects temporal flow variation and routing of flow through the watershed, collection system, and any storage facilities, the rational method should be used only for applications in which accuracy of runoff values is not essential. The empirical rational formula is expressed as

$$Q_p = \frac{CiA}{K_R} \tag{2-3}$$

where Q_p is peak runoff rate in cfs or m^3/s; C is a dimensionless runoff coefficient used as an adjustment for rainfall abstractions and is listed in Table 2-2 as a function of land use; i is the average rainfall intensity in in/hr or mm/hr for a duration equal to the time of concentration, or time required for water to travel from the most remote portion of the basin to the point of concern (i.e., inlet time) plus travel time in any contributing upstream sewers; A is the drainage area in ac or ha; and K_R is a conversion constant equal to 1.0 in U.S. customary units and 360 in S.I. units.

Time of concentration for the basin area can be computed using one of the formulas listed in Table 2-3. Once the time of concentration is known, the intensity in Equation 2-3 can be obtained from regional intensity-duration-frequency (IDF) curves (see Figure 2-1) for the design runoff frequency. Common practice is to design storm sewers for a two- to ten-year return frequency in residential areas and ten to 30 years for commercial regions. IDF curves are often available from local water management, highway or drainage districts, regulatory agencies, and weather bureaus, such as the National Oceanic and Atmospheric Administration. In the absence of existing data, however, IDF curves can be estimated using National Weather Service frequency distributions (Hershfield, 1961) along with a methodology proposed by Chen (1983).

Table 2-2: Runoff coefficients for 2 to 10 year return periods

Description of drainage area	Runoff coefficient
Business	
Downtown	0.70-0.95
Neighborhood	0.50-0.70
Residential	
Single-family	0.30-0.50
Multi-unit detached	0.40-0.60
Multi-unit attached	0.60-0.75
Suburban	0.25-0.40
Apartment dwelling	0.50-0.70
Industrial	
Light	0.50-0.80
Heavy	0.60-0.90
Parks and cemeteries	0.10-0.25
Railroad yards	0.20-0.35
Unimproved areas	0.10-0.30
Pavement	
Asphalt	0.70-0.95
Concrete	0.80-0.95
Brick	0.75-0.85
Roofs	0.75-0.95
Lawns	
Sandy soils	
Flat (2%)	0.05-0.10
Average (2 – 7 %)	0.10-0.15
Steep (≥ 7%)	0.15-0.20
Heavy soils	
Flat (2%)	0.13-0.17
Average (2 – 7 %)	0.18-0.22
Steep (≥ 7%)	0.25-0.35

Source: Adapted from ASCE (1992)

For nonhomogeneous drainage areas having variable land uses, a composite runoff coefficient, C_c, should be used in the rational formula. The composite coefficient is expressed as

$$C_c = \frac{\sum_{j=1}^{n} C_j A_j}{\sum_{j=1}^{n} A_j} \tag{2-4}$$

where A_j is the area for land use j; C_j is the dimensionless runoff coefficient for area j; and n is the total number of land covers. If Equation 2-4 is substituted into Equation 2-3, the rational formula can be rewritten as

$$Q_p = \frac{i\sum_{j=1}^{n} C_j A_j}{K_R} \tag{2-5}$$

Application of the rational method is valid for drainage areas less than 200 ac (80 ha), which typically have times of concentration of less than 20 minutes (ASCE, 1992).

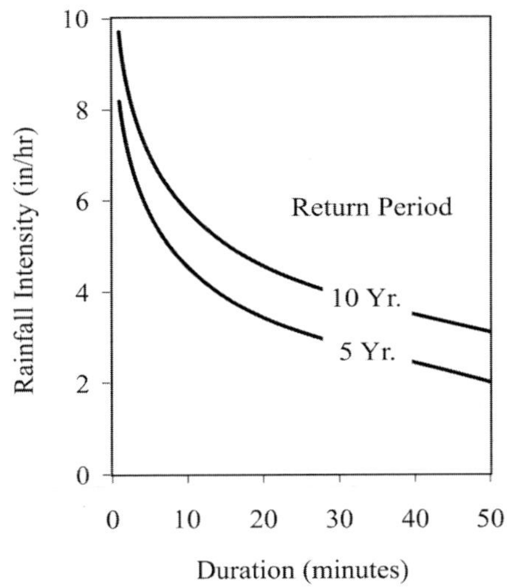

Figure 2-1: Sample IDF curves

2.3.2 Unit Hydrograph Methods

For larger areas where watershed or channel storage may be significant, the rational method is not appropriate for determination of wet weather loads. In these cases, it is necessary to evaluate the variation of flow over time, or the entire runoff hydrograph. In application, the hydrograph at the upstream end of a sewer can be used with various routing techniques to produce the outflow hydrograph at its downstream end. The simplest routing method involves lagging the hydrograph, without distortion, by the time required for flow to travel through the sewer. Then the combined outflow hydrograph for all upstream contributing mains, plus any additional surface runoff, represents the design inflow hydrograph to the adjacent downstream sewer.

Table 2-3: Formulas for computing time of concentration

Method	Formula	Comments
Kirpich (1940)	$t_c = 0.0078L^{0.77}S^{-0.385}$	For overland flow on concrete or asphalt, multiply t_c by 0.4; for concrete channels, multiply by 0.2
Izzard (1946)	$t_c = \frac{41.025(0.0007i + c)L^{1/3}}{S^{1/3}i^{2/3}}$	Retardance factor, c, ranges from 0.007 for smooth pavement to 0.012 for concrete and to 0.06 for dense turf; for $iL < 500$
FAA (1970)	$t_c = \frac{0.39(1.1 - C)L^{1/2}}{S^{1/3}}$	Runoff coefficient, C, from Table 2-2
Kinematic wave (Morgali and Linsley, 1965; Aron and Erborge, 1973)	$t_c = \frac{0.938L^{0.6}n^{0.6}}{i^{0.4}S^{0.3}}$	Manning roughness coefficient, n, found from Table 2-4
NRCS upland method (SCS, 1986)	$t_c = \frac{1}{60}\sum_{j=1}^{N}(L_j/V_j)$	For shallow concentrated or channel flow, average velocity, V, in segment j can be computed via Mannings equation; for overland flow, see NRCS charts (SCS, 1986) plotting V as a function of surface cover and slope
NRCS lag equation (SCS, 1986)	$t_c = \frac{100L^{0.8}[(1000/CN) - 9]^{0.7}}{19000S^{1/2}}$	Curve number, CN, is from Table 2-7
Yen and Chow (1983)	$t_c = K_Y\left(\frac{NL}{S^{1/2}}\right)^{0.6}$	K_Y ranges from 1.5 for light rain ($i < 0.8$) to 1.1 for moderate rain ($0.8 < i < 1.2$), and to 0.7 for heavy rain ($i > 1.2$); overland texture factor, N, in Table 2-5

Note: t_c is evaluated in minutes; L is length of the flow path in ft; i is rainfall intensity in in/hr; and S is average slope in ft/ft.

Table 2-4: Manning roughness coefficients, n, for overland flow

Surface description	Manning n
Concrete, asphalt	0.010 – 0.013
Bare sand	0.010 – 0.016
Gravel, bare clay-loam (eroded)	0.012 – 0.033
Natural rangeland	0.010 – 0.320
Bluegrass sod	0.390 – 0.630
Short-grass prairie	0.10 – 0.20
Dense grass, Bermuda grass, bluegrass	0.170 – 0.480
Forestland	0.20 – 0.80

Source: Adapted from Engman (1986)

A unit hydrograph is a linear conceptual model that can be used to transform rainfall excess into a runoff hydrograph. By definition, it is the hydrograph that results from 1 in, or 1 cm in S.I. units, of rainfall excess generated uniformly over the watershed at a uniform rate during a specified period of time. The process by which an existing unit hydrograph is used with given storm inputs to yield a direct runoff hydrograph is known as convolution. In discretized form, the convolution equation can be expressed as (Chow et al., 1988)

$$Q_n = \sum_{m=1}^{n \le M} P_m U_{n-m+1} \qquad for\ n = 1, 2, \ldots, N \tag{2-6}$$

where Q_n is a direct runoff hydrograph ordinate; P_m is excess rainfall at interval m; and U_{n-m+1} represents the unit hydrograph ordinate. Subscripts n and m designate the runoff hydrograph time interval and precipitation time interval, respectively.

Note that if total rainfall data (i.e., hyetograph) has been recorded, initial abstractions, such as interception storage and depression storage, and infiltration should be subtracted to define only the excess rainfall distribution. A popular means for estimating rainfall excess directly is the Natural Resources Conservation Service (NRCS) (formerly Soil Conservation Service) curve number method. The widely used TR-20 and TR-55 computer models (SCS, 1965), as well as many others, utilize the method to evaluate runoff peaks and volumes.

Table 2-5: Overland texture factor *N*

Overland flow surface	Low	Medium	High
Smooth asphalt pavement	0.010	0.012	0.015
Smooth impervious surface	0.011	0.013	0.015
Tar and sand pavement	0.012	0.014	0.016
Concrete pavement	0.014	0.017	0.020
Rough impervious surface	0.015	0.019	0.023
Smooth bare packed soil	0.017	0.021	0.025
Moderate bare packed soil	0.025	0.030	0.035
Rough bare packed soil	0.032	0.038	0.045
Gravel soil	0.025	0.032	0.045
Mowed poor grass	0.030	0.038	0.045
Average grass, closely clipped sod	0.040	0.050	0.060
Pasture	0.040	0.055	0.070
Timberland	0.060	0.090	0.120
Dense grass	0.060	0.090	0.120
Shrubs and bushes	0.080	0.120	0.180
Land use	**Low**	**Medium**	**High**
Business	0.014	0.022	0.035
Semi-business	0.022	0.035	0.050
Industrial	0.020	0.035	0.050
Dense residential	0.025	0.040	0.060
Suburban residential	0.030	0.055	0.080
Parks and lawns	0.040	0.075	0.120

Source: Yen and Chow (1983)

The curve number method separates total rainfall depth, P, into three components: depth of rainfall excess, P_e, initial abstractions, I_a, and retention, which consists primarily of the infiltrated volume of runoff. These components are related by (SCS, 1986)

$$P_e = \frac{(P - I_a)^2}{P - I_a + S} \tag{2-7}$$

where S is the potential maximum retention of the soil. From analysis of experimental watersheds,

$$I_a = 0.2S \tag{2-8}$$

so that Equation 2-7 can be expressed as

$$P_e = \frac{(P - 0.2S)^2}{P + 0.8S} \tag{2-9}$$

for $P > 0.2S$. Empirical studies by the NRCS indicate that the potential maximum retention can be estimated as

$$S = \frac{1000}{CN} - 10 \tag{2-10}$$

where CN represents a dimensionless runoff curve number between zero and 100 and is a function of land use, antecedent soil moisture, and other factors affecting runoff and retention. Soils are classified into four groups, A, B, C, and D, which are described in Table 2-6. Curve numbers for various land uses and soil groups are listed in Table 2-7. The curve number values provided, however, apply only to normal antecedent moisture conditions (AMCII). For AMCI (i.e., low moisture) or AMCIII (i.e., high moisture), the following approximations can be used to derive equivalent curve numbers:

$$CN(I) = \frac{4.2CN(II)}{10 - 0.058CN(II)} \tag{2-11}$$

$$CN(III) = \frac{23CN(II)}{10 + 0.13CN(II)} \tag{2-12}$$

For areas containing several subcatchments with differing curve numbers, an area-averaged composite CN can be computed. Note that the curve number method best represents a long-term expected relationship between rainfall and runoff and is not ideally suited for individual storms (Smith, 1997). In addition,

NRCS does not recommend the use of the curve number method when CN falls below a value of 40.

Table 2-6: Description of NRCS soil classifications

Group	Description	Min. infiltration (in/hr)
A	Deep sand; deep loess; aggregated silts	0.30 – 0.45
B	Shallow loess; sandy loam	0.15 – 0.30
C	Clay loams; shallow sandy loam; soils low in organic content; soils usually high in clay	0.05 – 0.15
D	Soils that swell significantly when wet; heavy plastic clays; certain saline soils	0 – 0.05

Source: SCS (1985)

2.3.2.1 Natural Unit Hydrograph

To develop a unit hydrograph from measured data, a gaged watershed ranging in size from 1.0 and 1,000 mi^2 (2.6 and 2,600 km^2) should be selected. Assuming that a sufficient number of rainfall-runoff records can be obtained for the watershed, selection of specific events to use in the analysis should be made in accordance with the following criteria (Viessman and Lewis, 1996):

- Storms should have a simple structure (i.e., individually occurring) with relatively uniform spatial and temporal rainfall distributions;
- Direct runoff should range from 0.5 to 1.75 in (1.25 to 4.5 cm);
- Duration of the rainfall event should range from 10 to 30 percent of the lag time, defined as the time from the midpoint of the excess rainfall to the peak discharge; and
- At least several storms that meet the previous criteria and that have a similar duration of excess rainfall should be analyzed to obtain average rainfall-runoff data.

Once data are selected, the measured time distribution of rainfall excess, P, and direct runoff ordinates, Q, are applied within a reverse convolution, or deconvolution, process to derive the unit hydrograph. Assuming there are M discrete values of excess rainfall that define a storm event and N discrete values of direct runoff, then from Equation 2-6, N equations can be written for Q_n, $n = 1, 2\ldots, N$, in terms of $N - M + 1$ unit hydrograph ordinates (Mays, 2001). For example,

$$\left\{\begin{array}{l} Q_1 = P_1 U_1 \\ Q_2 = P_2 U_1 + P_1 U_2 \\ \ldots \\ Q_M = P_M U_1 + P_{M-1} U_2 + \ldots + P_1 U_M \\ Q_{M+1} = 0 + P_M U_2 + \ldots + P_2 U_M + P_1 U_{M+1} \\ \ldots \\ Q_{N-1} = 0 + 0 + \ldots + 0 + 0 + \ldots + P_M U_{N-M} + P_{M-1} U_{N-M+1} \\ Q_N = 0 + 0 + \ldots + 0 + 0 + \ldots + 0 + P_M U_{N-M+1} \end{array}\right\} \qquad (2\text{-}13)$$

represents a set of N equations with $N - M + 1$ unknowns that can be solved algebraically or by matrix operators. As a final step, since averaged data was used in the analysis, the unit hydrograph should be adjusted to ensure that the distribution corresponds to 1 in, or 1 cm, of direct runoff.

2.3.2.2 Synthetic Unit Hydrograph

In the previous discussion, it was assumed that the design storm was applied to the same watershed from which the unit hydrograph was derived. In many applications, however, rainfall and runoff data are not available. Synthetic unit hydrograph procedures are thus used to develop unit hydrographs for ungaged locations in the watershed or for other watersheds that have similar runoff generation behavior considering characteristics such as geomorphology, soils, land cover/land use, and climate. Many synthetic unit hydrograph methods have been proposed in the hydrologic literature. Some of the most commonly used techniques are the NRCS dimensionless unit hydrograph method (SCS, 1985), the NRCS triangular unit hydrograph method (SCS, 1985), Snyder's method (Snyder, 1938), Clark's unit hydrograph method (Clark, 1945), the Colorado Urban Hydrograph Procedure (UDFCD, 2002), and the tri-triangular method (Boulos, 2004a-b).

NRCS Dimensionless Unit Hydrograph Method. In development of the NRCS dimensionless unit hydrograph, which is tabulated in Table 2-8 and illustrated in Figure 2-2, unit hydrographs from a large number of watersheds were evaluated, averaged, and made dimensionless.

The dimensionless time and runoff ordinates can then be dimensionalized by multiplying the corresponding values (t/t_p or Q/Q_p) by time from the beginning of excess rainfall to the time of peak discharge, t_p, or the peak runoff, Q_p, respectively.

Table 2-7: Runoff curve numbers for urban land uses

Land use description	Soil group			
	A	B	C	D
Lawns, open spaces, parks, golf courses:				
Good condition: grass cover on 75% or more of area	39	61	74	80
Fair condition: grass cover on 50% to 75% of area	49	69	79	84
Poor condition: grass cover on 50% or less of area	68	79	86	89
Paved parking lots, roofs, driveways, etc	98	98	98	98
Streets and roads:				
Paved with curbs and storm sewers	98	98	98	98
Gravel	76	85	89	91
Dirt	72	82	87	89
Paved with open ditches	83	89	92	93
Commercial and business areas (85% impervious)	89	92	94	95
Industrial districts (72% impervious)	81	88	91	93
Row houses, town houses and residential with lot sizes of 1/8 ac or less (65% impervious)	77	85	90	92
Residential average lot size:				
1/4 ac (38% impervious)	61	75	83	87
1/3 ac (30% impervious)	57	72	81	86
1/2 ac (25% impervious)	54	70	80	85
1 ac (20% impervious)	51	68	79	84
2 ac (12% impervious)	46	65	77	82
Developing urban area (newly graded; no vegetation)	77	86	91	94

Source: SCS (1985)

Based on NRCS recommendation, time to peak discharge can be estimated by

$$t_p = \frac{2}{3} t_c \qquad (2\text{-}14)$$

where t_c is the time of concentration for the basin area, which should be computed using one of the NRCS formulas listed in Table 2-3.

Table 2-8: NRCS dimensionless unit hydrograph

t/tp	Q/Qp	t/tp	Q/Qp
0.0	0.000	1.4	0.780
0.1	0.030	1.5	0.680
0.2	0.100	1.6	0.560
0.3	0.190	1.8	0.390
0.4	0.310	2.0	0.280
0.5	0.470	2.2	0.207
0.6	0.660	2.4	0.147
0.7	0.820	2.6	0.107
0.8	0.930	2.8	0.077
0.9	0.990	3.0	0.055
1.0	1.000	3.5	0.025
1.1	0.990	4.0	0.011
1.2	0.930	4.5	0.005
1.3	0.860	5.0	0.000

Source: SCS (1985)

Q_p in cfs/in or m^3/s/m is defined as

$$Q_p = \frac{K_p A}{t_p} \tag{2-15}$$

where A is the drainage area in mi^2 or km^2; K_p is a constant equal to 484 in U.S. customary units and 2.08 in S.I. units; and t_p is given in hours. The time associated with the recession limb of the unit hydrograph, or time from peak discharge to the end of direct runoff, can be approximated multiplying t_p by 1.67 for an equivalent triangular hydrograph or 4.0 for the curvilinear hydrograph.

The resulting synthetic unit hydrograph is applicable only for an effective duration of excess rainfall, t_r, recommended as (SCS, 1985)

$$t_r = 0.133 t_c \tag{2-16}$$

Depending on the application, the current duration of excess rainfall may not be convenient. For example, it is necessary to divide the design storm into a discrete number of time intervals. The duration that results from basin parameters will often not evenly divide into the design storm duration. In other cases, the effective duration may be of such magnitude that the number of computations can be reduced if a larger duration is utilized. Fortunately, the S-hydrograph method can be used to convert a unit hydrograph of any given duration into a unit hydrograph of any other desired effective duration. The S-hydrograph, or S-curve, theoretically represents the response of a particular watershed to constant rainfall excess for an indefinite period. It can be derived by adding an infinite series of lagged unit hydrographs, as shown in Figure 2-3. The S-hydrograph computed with the t_r duration unit hydrograph is lagged by the desired duration, t_r'. The difference between the two S-curves is then multiplied by the ratio t_r/ t_r'. The result is a new unit hydrograph having an effective duration of t_r'.

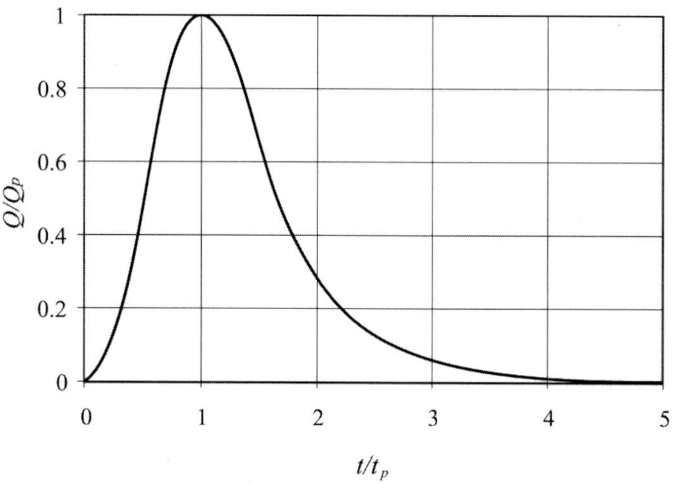

Figure 2-2: NRCS Dimensionless unit hydrograph (SCS, 1985)

NRCS Triangular Unit Hydrograph Method. The NRCS triangular unit hydrograph (Figure 2-4) is an approximation to the NRCS dimensionless unit hydrograph described above. The peak flow, the time to peak, and the effective rainfall duration are all determined using the same equations as for the dimensionless unit hydrograph. The attractive feature of the triangular unit hydrograph is its simplicity in the sense that the entire unit hydrograph is defined in terms of only three terms: the peak flow, the time to peak, and the time base. Unlike the dimensionless unit hydrograph that has time base of $5t_p$, the time base of the triangular unit hydrograph is $2.67t_p$.

Snyder Unit Hydrograph Method. Snyder's method for unit hydrograph synthesis relates the time from the centroid of the excess rainfall to the peak of the unit hydrograph, also referred to as lag time, to geometric characteristics of the basin in order to derive critical points for interpolating the unit hydrograph. Lag time is evaluated by

$$t_L = C_1 C_t \left(L L_{CA}\right)^{0.3} \qquad (2\text{-}17)$$

where t_L is in hrs; C_1 is a constant equal to 1.0 in U.S. customary units and 0.75 in S.I. units; C_t is an empirical watershed storage coefficient, which generally ranges from 1.8 to 2.2; L is the length of the main stream channel in mi or km; and L_{CA} is the length of stream channel from a point nearest the center of the basin to the outlet in mi or km.

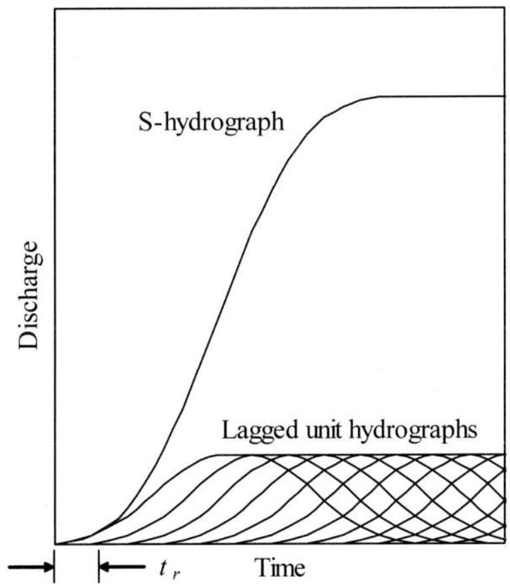

Figure 2-3: Development of an S-hydrograph

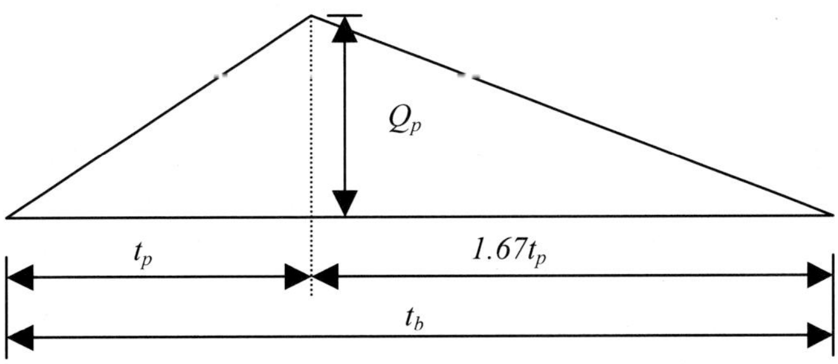

Figure 2-4: NRCS Triangular unit hydrograph (SCS, 1985)

The standard duration of excess rainfall is computed empirically by

$$t_r = \frac{t_L}{5.5} \qquad (2\text{-}18)$$

Adjusted values of lag time, t_{La}, for other durations of rainfall excess can be obtained by

$$t_{La} = t_L + 0.25(t_{ra} - t_r) \tag{2-19}$$

where t_{ra} is the alternative unit hydrograph duration. Time to peak discharge can be computed as a function of lag time and duration of excess rainfall, expressed as

$$t_p = t_{La} + 0.5t_{ra} \tag{2-20}$$

The peak discharge, Q_p is defined as

$$Q_p = \frac{C_2 C_p A}{t_{La}} \tag{2-21}$$

where Q_p is in cfs/in or $m^3/s/m$; C_2 is a constant equal to 640 in U.S. customary units and 2.75 in S.I. units; A is drainage area in mi^2 or km^2; and C_p is a second empirical constant ranging from approximately 0.5 to 0.7. Coefficients C_t and C_p are regional parameters that should be calibrated or be based on values obtained for similar gaged drainage areas. The ultimate shape of Snyder's unit hydrograph is primarily controlled by two parameters, W_{50} and W_{75}, which represents widths of the unit hydrograph at discharges equal to 50 and 75 percent of the peak discharge, respectively. These shape parameters can be evaluated by

$$W_{50} = C_{50}\left(\frac{A}{Q_p}\right)^{1.08} \tag{2-22}$$

and

$$W_{75} = C_{75}\left(\frac{A}{Q_p}\right)^{1.08} \tag{2-23}$$

where C_{50} is a constant equal to 770 in U.S. customary units and 2.14 in S.I. units; and C_{75} is a constant equal to 440 in U.S. customary units and 1.22 in S.I. units. The location of the end points for W_{50} and W_{75} are often placed such that one-third of both values occur prior to the time to peak discharge and the remaining two-thirds occur after the time to peak. Finally, the base time, or time from beginning to end of direct runoff, should be evaluated such that the unit hydrograph represents 1 in (or 1 cm in S.I. units) of direct runoff volume. With known values of t_p, Q_p, W_{50}, and W_{75}, along with the adjusted base time, one can then locate a total of seven unit hydrograph ordinates.

Clark Unit Hydrograph Method. Clark's method derives a unit hydrograph by explicitly representing the processes of translation and attenuation, which are the two critical phenomena in transformation of excess rainfall to runoff hydrograph. Translation refers to the movement, without storage, of runoff from its origin to the watershed outlet in response to gravity force, where as attenuation represents the reduction of runoff magnitude due to resistances arising from frictional forces and storage effects of soil, channel, and land surfaces. Clark (1945) noted that the translation of flow through the watershed could be described by a time-area curve (Figure 2-5), which expresses the curve of the fraction of watershed area contributing runoff to the watershed outlet as a function of travel time since the start of effective precipitation. Each subarea is delineated so that all the precipitation falling on the subarea instantaneously has the same time of travel to the outflow point.

Developing a time-area curve for a watershed could be a time consuming process. For watersheds that lack derived time-area diagram, the HEC-HMS model, which was developed at the Hydrologic Engineering Center (HEC) of the U.S. Army Corps of Engineers, uses the following relationship (HEC, 2000)

$$\frac{A_{c,t}}{A_T} = \begin{Bmatrix} 1.414\left(\frac{t}{t_c}\right)^{1.5} & for \ \ t \le \frac{t_c}{2} \\ 1 - 1.414\left(1 - \frac{t}{t_c}\right)^{1.5} & for \ \ t \ge \frac{t_c}{2} \end{Bmatrix} \qquad (2\text{-}24)$$

where $A_{c,t}$ is cumulative watershed area contributing at time t; A_T is total watershed area; and t_c is time of concentration of the watershed. If the incremental areas, denoted as A_i in Figure 2-5, are multiplied by a unit depth of excess rainfall and divided by Δt, the computational time step, the result is a translated hydrograph that is considered as an inflow to a conceptual linear reservoir located at the watershed outlet.

To account for storage effects, the attenuation process is modeled by routing the translated hydrograph through a linear reservoir with storage properties similar to those of the watershed. The routing model is based on the mass balance equation

$$\frac{dS}{dt} = I_t - Q_t \qquad (2\text{-}25)$$

where dS/dt is time rate of change of water in storage at time t; I_t is average inflow, obtained from the time-area curve, to storage at time t; and Q_t is outflow from storage at time t.

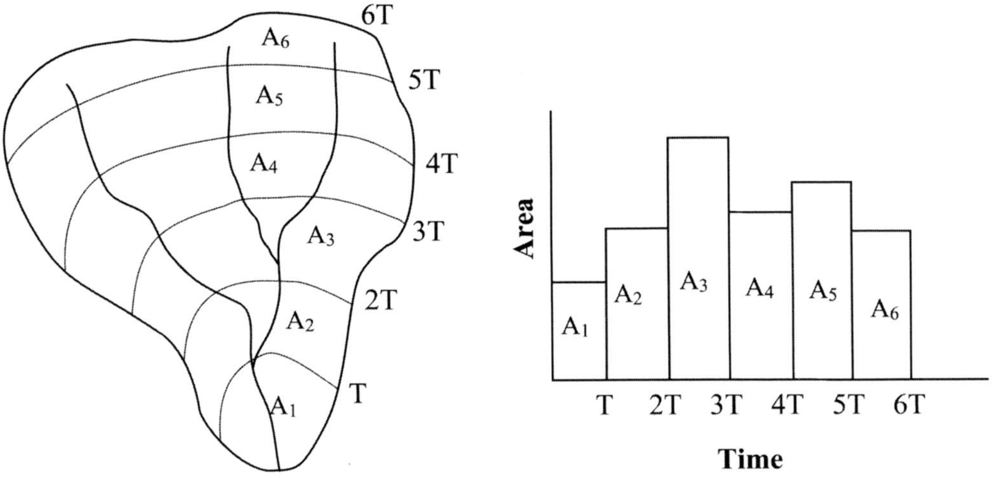

Figure 2-5: Time-area histogram for a watershed

For linear reservoir model, storage is related to outflow as

$$S_t = RQ_t \tag{2-26}$$

where R is a constant linear reservoir parameter that represents the storage effect of the watershed. Usually, lag time (t_L) is used as an approximation to R. Combining and solving Equations 2-25 and 2-26 using a finite difference approximation provides

$$Q_t = C_1 I_t + C_2 Q_{t-1} \tag{2-27}$$

where C_1 and C_2 are routing coefficients calculated as

$$C_1 = \frac{\Delta t}{R + 0.5\Delta t} \tag{2-28}$$

$$C_2 = 1 - C_1 \tag{2-29}$$

The average outflow during period t is

$$\bar{Q}_t = \frac{Q_{t-1} + Q_t}{2} \tag{2-30}$$

If the inflow, I_t, ordinates are runoff from a unit depth of excess rainfall, the average outflows derived by Equation 2-30 represent Clark's unit hydrograph ordinates. Clark's unit hydrograph is, therefore, obtained by routing a unit depth of direct runoff to the channel in proportion to the time-area curve and routing the runoff entering the channel through a linear reservoir. Note that solution of Equations 2-27 and 2-30 is a recursive process. As such, average outflow ordinates of the unit hydrograph will theoretically continue for an infinite duration. Therefore, it is customary to truncate the recession limb of the unit hydrograph where the outflow volume exceeds 0.995 inches or mm. Clark's method is based on the premise that duration of the rainfall excess is infinitesimally small. Because of this, Clark's unit hydrograph is referred to as an instantaneous unit hydrograph or IUH. In practical applications, it is usually necessary to alter the IUH into a unit hydrograph of specific duration. This can be accomplished by lagging the IUH by the desired duration and averaging the ordinates.

Colorado Urban Hydrograph Procedure. The Colorado Urban Hydrograph Procedure (CUHP) is an adaptation of Snyder's method based on data for Colorado urban watersheds ranging in size from 100-200 acres (UDFCD, 1984). The technique is most commonly used in the state of Colorado to derive a unit hydrograph for urban and rural watersheds that have areas ranging from 90 acres to 5 square miles. Whenever a larger watershed is studied, it is recommended to subdivide the watershed into subcatchments of 5 square miles or less. The shape of the CUHP unit hydrograph (Figure 2-6) is determined using the empirical equations presented below. These equations relate unit hydrograph parameters to physical characteristics of the watershed. The method considers the effects of watershed size, shape, percentage of the total surface area that is impervious, length of the main drainage channel, slope, and other essential watershed behavior.

Lag time (t_L) of the watershed, defined as the time from the center of unit storm duration to the peak of the unit hydrograph, is determined as

$$t_L = C_t \left(\frac{L \cdot L_{Ca}}{\sqrt{S}} \right)^{0.48} \tag{2-31}$$

where t_L is in hours; L is length along the drainageway path from study point to the most upstream limits of the catchment in miles; L_{ca} is length along stream from study point to a point along stream adjacent to the centroid of the catchment in miles; S is length weighted average slope of catchment along drainageway path to upstream limits of the catchment; and C_t is time to peak coeficient. Once the lag time is determined, the time to peak (t_p) of the unit hydrograph could be obtained by adding $0.5t_r$ to the lag time in consistent units.

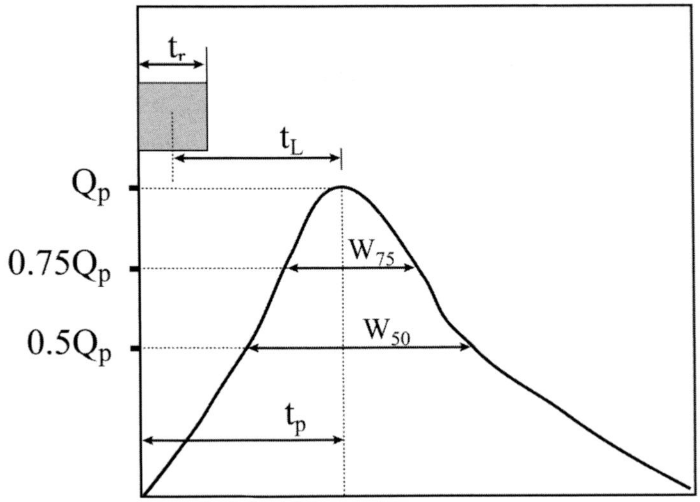

Figure 2-6: The CUHP unit hydrograph

Peak flow rate, Q_p, of the unit hydrograph is calculated as

$$Q_p = \frac{640 C_p A}{t_p} \tag{2-32}$$

where Q_p is peak flow rate of the unit hydrograph, in cfs; A is area of the catchment, in square miles; C_p is unit hydrograph peaking coefficient, and is determined as

$$C_p = P \cdot C_t \cdot A^{0.15} \tag{2-33}$$

where P is peaking parameter. C_t and P are defined in terms of percent impervious (I_a) of the catchment as

$$C_t = aI_a^2 + bI_a + c \tag{2-34}$$

$$P = dI_a^2 + eI_a + f \tag{2-35}$$

The coefficients a, b, c, d, e, and f are defined in terms of I_a in Table 2-9. The capability of the CUHP to account for percent imperviousness of the watershed to derive a synthetic unit hydrograph makes it the method of choice for urban watersheds.

Table 2-9: CUHP coefficients as a function of percent imperviousness

I_a	a	b	c	d	e	F
$I_a \leq 10$	0.0	-0.00371	0.163	0.00245	-0.012	2.16
$10 \leq I_a \leq 40$	2.3×10^{-5}	-0.00224	0.146	0.00245	-0.012	2.16
$I_a \geq 40$	3.3×10^{-5}	-0.000801	0.120	-0.00091	0.228	-2.06

The widths of the unit hydrograph at 50% and 75% of the peak are estimated as

$$W_{50} = \frac{500}{\left(\frac{Q_p}{A}\right)} \tag{2-36}$$

$$W_{75} = \frac{260}{\left(\frac{Q_p}{A}\right)} \tag{2-37}$$

where W_{50} is width of the unit hydrograph at 50 percent of the peak, in hours; W_{75} is width of the unit hydrograph at 75 precent of the peak, in hours; Q_p is peak flow rate, in cfs; and A is catchment area, in square miles. In addition to knowing the location of the unit hydrograph peak, and W_{50} and W_{75}, it also helps to know how to distribute the two widths around the peak. As a general rule, the smaller of 35 percent of W_{50} and $0.6t_p$ is assigned to the left of the peak at 50 percent of the peak, and 65 percent of W_{50} is assigned to the right of the peak. The width assigned to the left side of the peak at 75 percent of the peak depends on the case used for allocation of W_{50} to the left side of the peak at 50 percent of the peak. If 35 percent of W_{50} is assigned to the left at 50 percent of the peak, then 45 percent of W_{75} is given to the left side at 75 percent of the peak. Otherwise, left width at 75 percent of the peak will be $0.424t_p$. Right side of the peak is always equal to 55 percent of W_{75}.

Tri-triangular Unit Hydrograph Method. The tri-triangular method (Figure 2-7) is commonly used to derive rainfall dependent inflow/infiltration (RDII) flows for sewer collection systems. The technique applies up to three triangular hydrographs, as the name implies, to derive a unit hydrograph. The synthetic hydrograph is obtained by adding corresponding ordinates of the three triangular hydrographs. Each of these three triangular hydrographs has its own characteristic parameters, namely time to peak, recession constant, and fraction of an effective rainfall volume allocated to the triangle. R_1, R_2, and R_3 are fractions of excess rainfall volume, R, allocated to triangular hydrographs 1, 2, and 3 respectively. T_i and K_i are time to peak and recession constants of the triangles, respectively.

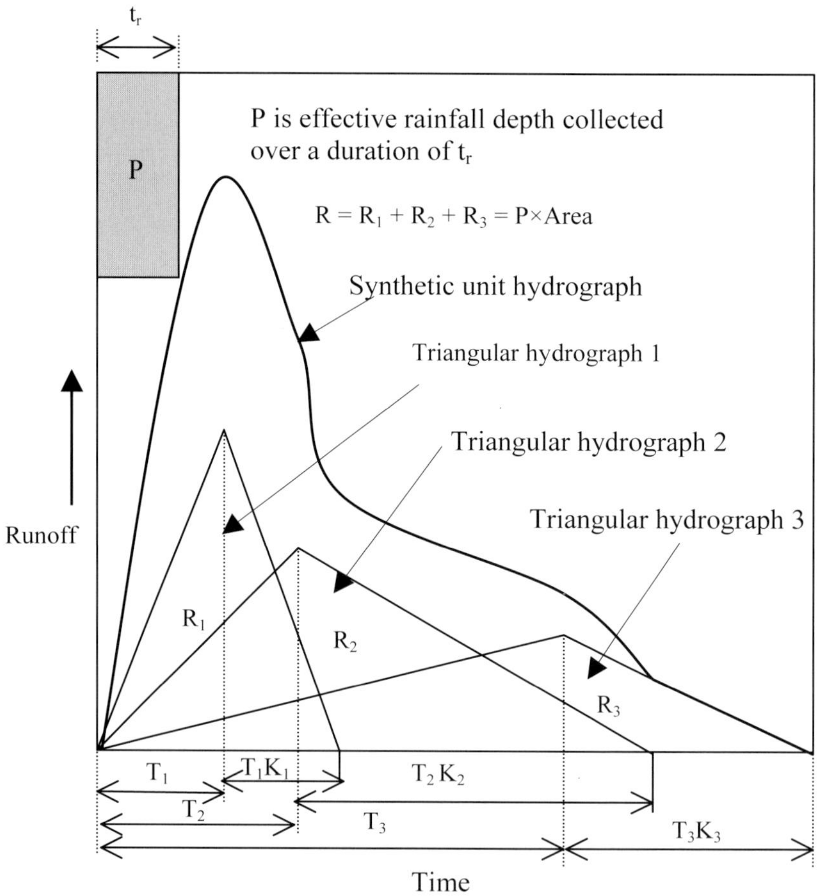

Figure 2-7: The tri-triangular unit hydrograph

The three triangular hydrographs are conceptual representations of different components of direct runoff or RDII. The first triangle represents rapidly responding (fast) components, such as contributions from pavements and rooftops, or direct inflow or rapid infiltration into separate sewer systems. The third triangle represents slow runoff components such as ground water contributions or slow infiltration into sewers. The second triangle represents runoff or infiltration with a medium time response. Time to peak value of the first triangle typically varies between 1 and 2 hours, depending on the size of the tributary area in question. The second triangle takes *T* values ranging from 4 to 8 hours. The third triangle parameter varies greatly depending on the infiltration characteristics of the system being modeled, and has a *T* value generally between 10 and 24 hours. The value of *K* for the first triangle typically ranges between 2 and 3. The second and third triangles assume *K* values from 2 to 4.

2.3.3 Physically-Based Models

Unit hydrograph methods are essentially empirical approaches for runoff computation that circumvent the need to solve advanced equations that govern various components of the hydrologic cycle (e.g., the St. Venant equations for surface flow routing, Richard's equations for flow routing in porous media). From a practical perspective, these approaches may be well justified: (1) The various components of runoff generation and flow are not entirely understood; and (2) the complexity of processes and the various solution techniques make manual solution techniques or coding of computational schemes impractical for the average practicing engineer (Westphal, 2001). Particularly for cases in which more advanced approaches may be warranted, the engineer may turn to hydrologic simulation software packages.

Over the last three decades, a number of computer-based hydrologic simulation models have been developed to simulate rainfall-runoff processes. They vary significantly in degree of complexity and data requirements. Singh and Woolhiser (2002) provide a comprehensive list and discussion of the numerous existing models. Physically-based hydrologic models, a particular class of models, are based on an understanding of the physics of the hydrologic processes that control watershed response and use related equations (e.g., St. Venant equations) to describe these physical processes. As a result, such models are far more adaptable and powerful than empirical techniques. Physically-based models are also generally categorized as continuous, distributed models, indicating an ability to capture both spatial and long-term temporal variability of basin response by accounting for all runoff components and emphasizing an overall moisture balance within the basin. At one time, such models were considered to be too computationally and data intensive to use for projects other than major research undertakings. However, physically-based models are more commonly being disseminated in versions compatible with personal computers and are being adapted for user-friendly interface, simplified data input, and graphical display of output.

2.3.3.1 Overview

While a number of physically-based models exist, the following paragraphs provide a brief summary of some of the more commonly used models.

CASC2D. Developed at the Center for Excellence in Geosciences at Colorado State University, the Cascade Two Dimensional Model (CASC2D) (Julien and Saghafian, 1991; Ogden, 1998) is one of the most advanced physically-based models available today. It solves the complete conservation equations for mass, energy, and linear momentum at a user-specified spatial resolution and at short time intervals (i.e., 1 to 30 seconds). Since its original development, the model has been significantly enhanced under funding from the U.S. Army Research Office and U.S. Army Corps of Engineers. Some notable features of the model include continuous accounting of soil-moisture, simulation of rainfall interception, retention, and infiltration, routing of surface runoff, and watershed-scale sediment transport simulation. CASC2D is capable of both continuous (i.e., long term) and single-event analysis.

DWSM. The Illinois State Water Survey's Dynamic Watershed Simulation Model (DWSM) (Borah, 1999) was developed to simulate propagation of flood waves and the entrainment and transport of sediment and commonly used agricultural chemicals for rural watersheds. It is a single-event model that can be applied to large watersheds due to integration of robust algorithms and solution techniques. Nonuniformities in topography, soil, and land use data are handled by dividing the watershed into sub-watersheds, or more specifically, one-dimensional overland, channel, and reservoir flow elements. Spatial distribution of rainfall is handled by assigning different breakpoint rainfall records to each overland flow segment. For runoff evaluation, DWSM's hydrologic module uses analytical and approximate analytical solutions to the kinematic wave equations, which include continuity and a simplified form of the conservation of momentum equation in which pressure and inertial forces are neglected (see Chapter 4).

HEC-HMS. The U.S. Army Corps of Engineers Hydrologic Engineering Center's (HEC) HEC-HMS (Hydrologic Modeling System) (HEC, 2000) has evolved over time from the popular HEC-1 (HEC, 1990) runoff model. HEC-HMS provides several methods for computing infiltration losses and for transforming excess precipitation into runoff. Runoff can be evaluated by the kinematic wave method with multiple horizontal planes or with simpler and empirically-based (i.e., hydrograph) methods. In addition, a meteorological module currently offers seven different historical and synthetic methods for generation of precipitation data and one method for evapotranspiration analysis. A flexible optimization tool is also provided to allow for convenient calibration of model parameters. Although it has many of the same capabilities as HEC-1, important differences are that it allows continuous simulation over long periods and distributed runoff computation using a grid cell depiction of the watershed, and it offers improved user interface and reporting capabilities. In particular, the model takes advantage of geospatial data provided through Geographic Information Systems (GIS) or Computer Aided Design and Drafting (CADD) programs.

KINEROS. The U.S. Department of Agriculture Agricultural Research Service developed the KINematic runoff and EROSion model referred to as KINEROS (Woolhiser et al., 1990) to simulate processes of interception, infiltration, surface runoff, and erosion from small agricultural and urban basins. The model is event-oriented since the model does not describe evapotranspiration and soil water movement, and thus a hydrologic balance, between storms. KINEROS uses a kinematic wave approximation of overland and channel flows

PRMS – Storm Mode. The Precipitation-Runoff Modeling System (PRMS) (Leavesley et al., 1983) was developed by the U.S. Geological Survey (USGS) to simulate basin response over long periods. Basins are divided into homogeneous spatial units called Hydrologic Response Units (HRUs), which are defined based on factors such as surface slope, aspect, elevation, soil type, vegetation, and rainfall distribution. Water and energy balances are computed daily for each HRU, made up of one or more interconnected flow planes, and the sum of responses for each HRU yields the daily basin response. In storm mode, the model provides simulations using variable time steps as small as one minute, and the second level of basin subdivision is used to permit evaluation of short-term response. PRMS has recently been added to the Modular Modeling System (MMS) as part of a Watershed Modeling Systems Initiative undertaken by USGS and the U.S. Bureau of Reclamation. The MMS makes use of a variety of compatible modules for simulating water, energy, and biogeochemical processes, and offers a GIS interface that helps in visualizing model parameters.

SWMM. The U.S. Environmental Protection Agency's Storm Water Management Model (SWMM) (Huber and Dickinson, 1988), originally developed in the 1970's and completely re-written in 2004 (Rossman, 2004), was designed for continuous or event-based simulation of subcatchements, conveyance, storage, treatment and receiving streams. The model considers both water quantity and quality, and flow routing can be performed using the kinematic wave method or with the full St. Venant equations.

2.3.3.2 Limitations

Simulation inherently involves mathematical abstraction of real systems, and therefore, some degree of system misrepresentation is likely to occur. In any modeling application, it is the responsibility of the modeler to carefully assess and interpret the results in light of study objectives, quality of data (e.g., potentially uncertain or incomplete inputs), and limits or errors imposed through a particular model. Given the complexity of physically-based model formulations, assessment can be difficult.

Physically-based models require a significant amount of rainfall/runoff data for proper calibration. Except for some larger urban areas, however, sufficient and reliable data are not available. Moreover, in areas where such data exist, they are rarely available for the specific basin being analyzed. Without an ability to properly calibrate, it is difficult to assess the accuracy of the hydrologic method or model in computing runoff. A partial, but by no means complete, solution to this problem involves estimating runoff using several different methods and comparing results. While such an approach does not ensure accuracy, it can promote consistency and some degree of confidence in results.

Physically-based models also suffer from problems of scale. While field measurements are typically taken at the point or local scale, actual model applications are at much larger scales. The variability, particularly evident in characterization of soil properties and precipitation, will typically increase with basin size. To date, however, broader and generalizable effects of spatial scale are not well defined.

2.4 SOLVED PROBLEMS

Problem 2.1 Dry weather load

An existing sanitary sewer system serves a residential and commercial area consisting of the following components:

- Low rise apartments housing 75 persons
- Single-family homes housing 125 persons
- One hotel with 30 rooms, each with a private bath
- Two restaurants, each serving two meals per day and having a 75 person capacity
- One gas station serving a maximum of 100 vehicles per day

Based on an evaluation of actual flow records, peak load factors of 2.5 and 1.8 apply for residential and commercial flows, respectively. Estimate the current daily average and peak domestic wastewater flows.

Solution

Estimated flows from each component can be established using data provided in Table 2-1.

- Apartment: *60 gpd/person × 75 persons = 4500 gpd*

- Single-family homes: *75 gpd/person × 125 persons = 9375 gpd*
- Hotel: *60 gpd/guest × 2 guests/room × 30 rooms = 3600 gpd*
- Restaurants: *8 gpd/patron × 75 patrons/meal × 2 meals/estab. × 2 estab. = 2400 gpd*
- Gas station: *10 gpd/vehicle × 100 vehicles = 1000 gpd*

The sum of average loads for each component comprises the total average flow, Q_{avg}.

$$Q_{avg} = (4500 + 9375 + 3600 + 2400 + 1000) = \underline{20,875\ gpd}$$

Multiplying the averages by their corresponding peak load factors yields peak flow, Q_p.

$$Q_p = (4500 + 9375)(2.5) + (3600 + 2400 + 1000)(1.8) = \underline{47,290\ gpd}$$

Comments: For assessment of sewer design capacity, an estimate of infiltration should be included, and figures should be viewed in light of expected growth and facility design life.

Problem 2.2 Sewer infiltration rate

Average wastewater flow for a small city is 300,000 gpd during the dry period of the year, when rainfall is minimal and groundwater levels are low. During the wet season, however, flows average 520,000 gpd. The sewer consists of two miles of 6-in diameter pipe, three miles of 10-in diameter pipe, and two miles of 15-in diameter pipe. Based on measured flows before and during a recent storm, the maximum hourly flow was 1.1 Mgpd during the storm and 840,000 gpd for the preceding period. Evaluate and comment on the infiltration rate.

Solution

Since infiltration is expected to be minimal during dry weather conditions, the average infiltration rate can be evaluated as the difference between average dry and wet weather averages.

$$Average\ infiltration = 520,000 - 300,000 = \underline{220,000\ gpd}$$

The maximum hourly infiltration rate is taken as the difference between peak hourly flows for wet and dry periods.

$$Maximum\ hourly\ rate = 1.1 \times 10^6 - 840,000 = \underline{260,000\ gpd}$$

The unit infiltration rate can be evaluated considering the composite diameter-length of sewers and the average infiltration rate.

$$Composite\ diameter\ length = (6\ in \times 2\ mi) + (10\ in \times 3\ mi) + (15\ in \times 2\ mi) = 72\ in\text{-}miles$$

$$Unit\ infiltration\ rate = \frac{220{,}000}{72} = \underline{\underline{3{,}055\ \frac{gpd}{in - mile}}}$$

Comments: Based on the unit rate, average infiltration may not seem excessive. However, the peak flow during the storm is more than 350 percent of the dry weather average, which would require oversizing of facilities. Methods to decrease total hydraulic load on the sewer and associated components should be investigated to minimize treatment costs.

Problem 2.3 Peak flow calculation

The sample sanitary sewer system shown in Figure P2-3a comprises 5 pipe sections, 5 manholes and one downstream treatment plant. The loading at each manhole (junction) is shown on the figure. Determine the peak flow in each pipe and the total flow entering the treatment plant. Use Equation 2-1 with K = 2.4 and ρ = 0.89.

Solution

The flow in each pipe segment is peaked based on the total (accumulated) flow contribution of upstream manholes. The resulting flows are depicted in Figure P2-3b.

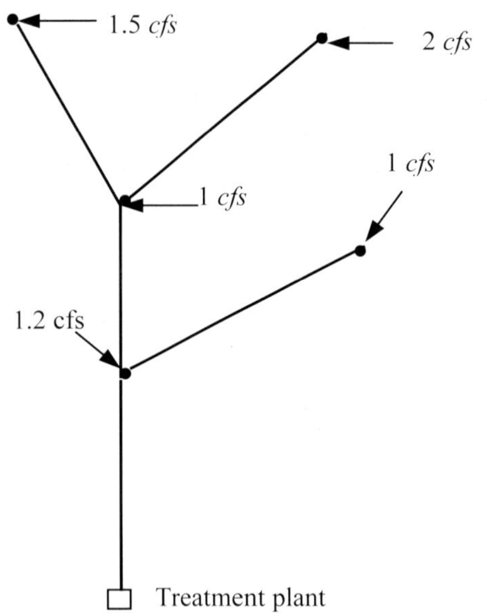

Figure P2-3a

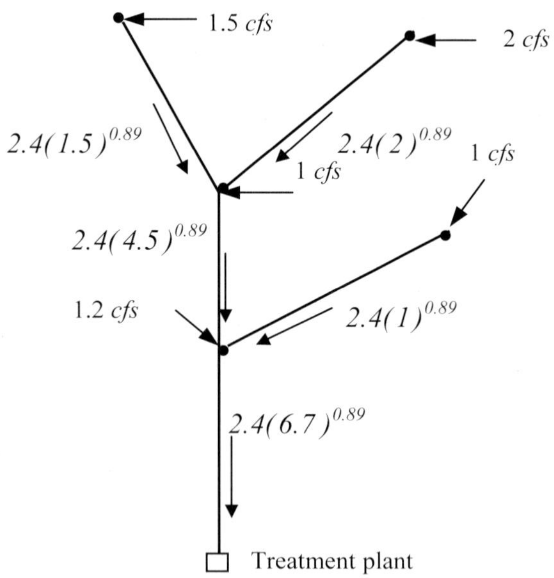

Figure P2-3b

Problem 2.4 Rational method

A new 7-ac suburban development is to be drained by a storm sewer that connects to a municipal drainage system. The time of concentration for the basin is 20 min, and the local IDF relationship can be approximated as ($i = 5.6 - 0.2t_r$), where i is design rainfall intensity in in/hr, and t_r is rainfall duration in hrs. The development is characterized as two subbasins; one has a drainage area of five acres and a runoff coefficient of 0.4, while the other drains two acres and has a runoff coefficient of 0.7. For the given characteristics, determine the peak runoff. Compare the answer to that if the entire drainage basin is used with a median runoff coefficient of 0.55.

Solution

The composite runoff coefficient for the two subcatchments is computed from Equation 2-4 as

$$C_c = \frac{(0.4 \times 5) + (0.7 \times 2)}{7} = 0.49$$

Rainfall intensity can be computed using the given IDF relationship

$$i = 5.6 - 0.2\left(\frac{20}{60}\right) = 5.53\,\frac{in}{hr}$$

Peak discharge is obtained by applying the rational equation (Equation 2-3) with the composite runoff coefficient.

$$Q_p = \frac{(0.49)(5.53)(7)}{1.0} = \underline{\underline{19.0\ cfs}}$$

Comments: If the median runoff coefficient is used, computed peak discharge will increase to 21.3 cfs, representing a 12 percent increase in estimated design flows. Thus, inclusion of a proper composite coefficient prevents unnecessary oversizing of drainage facilities.

Problem 2.5 Rational method

A storm sewer system drains five subcatchments as shown in Figure P2-5 and described in Table P2-5a. Assume that the IDF relationship in Figure 2-1 applies. If the average flow velocity is 5 fps throughout the system, determine the 10-yr design flow for each 400-ft length of sewer.

Solution

Compute the flow time, t_f, associated with each sewer by dividing flow length by its corresponding average velocity.

$$t_f = \frac{L}{V} = \frac{400}{5} = 80\ sec = 1.33\ min$$

Table 2-5b shows the computations leading to the peak discharge, Q_p, for each sewer. Note that each sewer is designated by its upstream manhole identification.

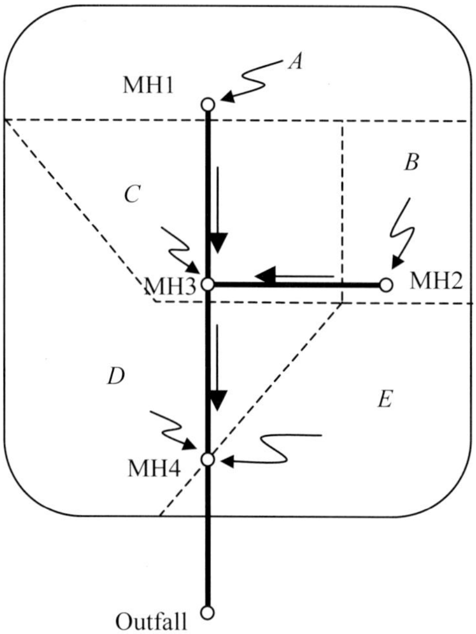

Figure P2-5

Table P2-5a

Catchment I.D.	Area (ac)	Runoff coefficient	Inlet time (min)
A	10	0.80	11.0
B	8	0.70	8.0
C	12	0.80	12.0
D	20	0.70	18.0
E	12	0.95	10.0

Table P2-5b

(1)	(2)	(3)	(4)	(5)	(6)	(7)	(8)	(9)	(10)	(11)	(12)
MH	A (ac)	*C*	*CA*	Σ*CA*	Flow path	t_i (min)	t_f (min)	t_c (min)	t_d (min)	*i* (in/hr)	Q_p (cfs)
1	10	0.80	8.0	8.0	A-1	11.0	-	11.0	11.0	5.6	**44.8**
2	8	0.70	5.6	5.6	B-2	8.0	-	8.0	8.0	6.2	**34.7**
					C-3	12.0	-	12.0			
3	12	0.80	9.6	23.2	A-1-3	11.0	1.33	12.33	12.33	5.45	**126.4**
					B-2-3	8.0	1.33	9.33			
	20	0.7	14.0	37.2	D-4	18.0	-	18.0			
4					E-4	10.0	-	10.0	18	4.8	**233.3**
	12	0.95	11.4	48.6	A-1-3-4	12.33	1.33	13.67			

Specific entries in the Table are as follows:

Column (1): The manhole that drains a particular subcatchment and the downstream sewer to which Q_p applies (e.g., Manhole no. 4 drains subcatchments *D* and *E* and simultaneously refers to the pipe leading to the outfall)

Column (2): Area of each subcatchment

Column (3): Value of the runoff coefficient for each subcatchment

Column (4): Product of *C* and the corresponding subcatchment area

Column (5): Summation of *CA* for all subcatchments drained by the sewer, which is equivalent to the sum of contributing previous values from Column 5 and the new value in Column 4 (e.g., for Manhole 3, 23.2 = 8.0 + 5.6 + 9.6)

Column (6): Identification of the flow paths being considered

Column (7): Values of inlet time, t_i, (see Table P2-5a)

Column (8): Upstream sewer flow time, t_f, for each flow path

Column (9): Summation of inlet and flow times for each flow path is the time of concentration, t_c

Column (10): For each manhole, the rainfall duration is longest time of concentration of different flow paths to arrive at the entrance of the sewer being considered; this value corresponds to the underlined value from Column 9

Column (11): The rainfall duration from Figure 2-1 that corresponds to a duration given in Column 10 and to a ten-yr frequency

Column (12): The design discharge for each of the four sewers computed from the rational equation (Equation 2-5)

Problem 2.6 Time of concentration (Kirpich)

The principle flow path for a 400-ac urban drainage basin is shown in Figure P2-6. Rainfall intensity, i, (in/hr) is expressed in mathematical form as $\frac{1.85}{(0.285+t_r)}$, where t_r is the rainfall duration in hrs. If the average flow velocity through storm sewer BC is 4 fps, determine the time of concentration at outlet C using the Kirpich equation.

Solution

Time of concentration, t_c, is equivalent to the sum of inlet time, t_i, and sewer flow time, t_f, where

$$t_f = \frac{2000}{4(60)} = 8.33\ min$$

Applying the Kirpich equation and adding sewer flow time,

$$t_i = 0.0078(1500)^{0.77}(0.018)^{-0.385} = 10.22\ min$$

$$t_c = 10.22 + 8.33 = \underline{\underline{18.55\ min}}$$

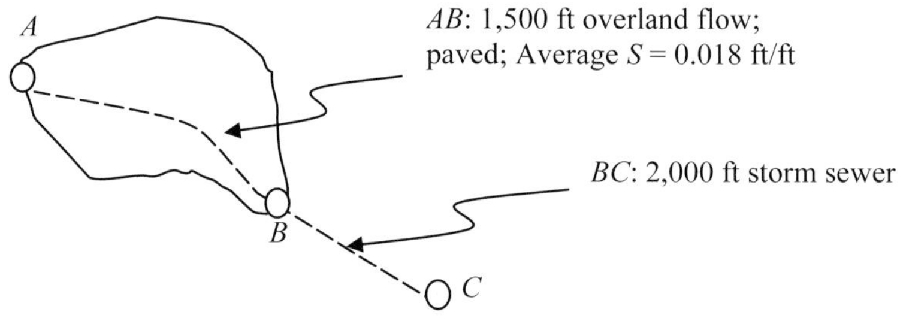

Figure P2-6

Problem 2.7 Time of concentration (Izzard)

Solve Problem 2.6 using the Izzard equation for computing time of concentration.

Solution

Assuming a conservative duration of 15 minutes,

$$i = \frac{1.85}{\left(0.285 + \frac{15}{60}\right)} = 3.46 \frac{in}{hr}$$

Note that iL = (3.46)(1500) = 5190, which is greater than the value of 500 recommended for use of the equation. Therefore, Izzard's equation does not apply to this basin/storm event.

Problem 2.8 Time of concentration (FAA)

Solve Problem 2.6 using the FAA equation for computing time of concentration.

Solution

Assuming C = 0.9 for application of the FAA equation and adding sewer flow time,

$$t_i = \frac{0.39(1.1 - 0.9)(1500)^{1/2}}{(0.018)^{1/3}} = 11.53\ min$$

$$t_c = 11.53 + 8.33 = \underline{\underline{19.86\ min}}$$

Problem 2.9 Time of concentration (kinematic wave)

Solve Problem 2.6 using the kinematic wave equation for computing time of concentration.

Solution

Assuming n = 0.013 and i = 3.46 in/hr,

$$t_i = \frac{0.938(1500)^{0.6}(0.013)^{0.6}}{(3.46)^{0.4}(0.018)^{0.3}} = 11.32\ min$$

Checking assumed rainfall intensity, for t_i = 11.32 min,

$$i = \frac{1.85}{\left(0.285 + \frac{11.32}{60}\right)} = 3.91 \frac{in}{hr}$$

Recompute the inlet time, based on the newly-computed rainfall intensity until values of i converge. The following summarizes the iterative solution, which yields a value of t_i equal to 10.70 min.

Assumed i (in/hr)	t_i (min)	Computed i (in/hr)
3.46	11.32	3.91
3.91	10.78	3.98
3.98	10.71	3.99
3.99	**10.70**	3.99 (OK)

Then, with sewer flow time,

$$t_c = 10.70 + 8.33 = \underline{\underline{19.03\ min}}$$

Problem 2.10 Time of concentration (NRCS lag)

Solve Problem 2.6 using the NRCS lag equation for computing time of concentration.

Solution

Assuming $CN = 98$ for application of the NRCS lag equation and adding sewer flow time,

$$t_i = \frac{100(1500)^{0.8}\left[\left(\frac{1000}{98}\right) - 9\right]^{0.7}}{19000(0.018)^{1/2}} = 15.52\ min$$

$$t_c = 15.52 + 8.33 = \underline{\underline{23.85\ min}}$$

Problem 2.11 Time of concentration (Yen and Chow)

Solve Problem 2.6 using the Yen and Chow equation for computing time of concentration.

Solution

Assuming $K_Y = 0.7$ and $N = 0.012$ for the Yen and Chow equation and adding sewer flow time,

$$t_i = 0.7\left(\frac{(0.012)(1500)}{(0.018)^{1/2}}\right)^{0.6} = 13.23\ min$$

$t_c = 13.23 + 8.33 = \underline{\underline{21.56\ min}}$

Comments: Computed time of concentration values range from 18.55 min. to 23.85 min, depending on the method of computation.

Problem 2.12 Convolution

Given the rainfall excess and 1-hr unit hydrograph (UH) below, determine the direct runoff hydrograph from the watershed. Assume a constant 0.3 in/hr rate of abstractions.

Time (hr)	1	2	3	4	5	6	7	8
Intensity (in)	0.5	1.0	1.7	0.5	-	-	-	-
UH (cfs/in)	100	320	450	370	250	160	90	40

Solution

The number of rainfall excess intervals, M, is equal to four. Substracting abstractions from total rainfall yields four 1-hr rainfall pulses as follows: $P_1 =$ 0.2 in, $P_2 = 0.7$ in, $P_3 = 1.4$ in, and $P_4 = 0.2$ in. In addition, there are eight unit hydrograph ordinates, so $N - M + 1 = 8$, and the number of direct runoff hydrograph ordinates, N, will be $8 + M - 1 = 11$. Applying Equation 2-6 to the first time interval, $n = 1$, runoff is evaluated as

$$Q_1 = P_1U_1 = (0.2)(100) = 20\ cfs$$

For the second and third intervals, $n = 2$ and 3,

$$Q_2 = P_1U_2 + P_2U_1 = (0.2)(320) + (0.7)(100) = 134\ cfs$$

and

$$Q_3 = P_1U_3 + P_2U_2 + P_3U_1 = (0.2)(450) + (0.7)(320) + (1.4)(100) = 454\ cfs$$

Formulation of similar equations will continue until $n = N = 11$. Referring to the summary of computations in Table P2-12, note that Column 3 shows the direct runoff hydrograph resulting from $P_1 = 0.2$ in; Column 4 shows the direct runoff from $P_2 = 0.7$ in; etc. Column 7 shows the total direct runoff hydrograph from the cumulative rainfall event. Each ordinate in the column is equivalent to the sum of ordinates across each row.

Problem 2.13 NRCS curve number method

Determine the rainfall excess for successive hourly periods for the following storm. Assume that the watershed is characterized by a curve number of 80.

Time (hr)	1	2	3	4	5	6	7
Intensity (in)	0.3	0.5	0.7	0.4	0.6	0.5	0.4

Table P2-12

(1)	(2)	(3)	(4)	(5)	(6)	(7)
		Total rainfall (in)				
Time, n (hr)	Unit hydrograph ordinates (cfs/in)	0.5	1.0	1.7	0.5	Direct runoff (cfs)
		Excess rainfall (in)				
		0.2	0.7	1.4	0.2	
0	0	0	-	-	-	**0**
1	100	20	0	-	-	**20**
2	320	64	70	0	-	**134**
3	450	90	224	140	0	**454**
4	370	74	315	448	20	**857**
5	250	50	259	630	64	**1003**
6	160	32	175	518	90	**815**
7	90	18	112	350	74	**554**
8	40	8	63	224	50	**345**
9	0	0	28	126	32	**186**
10	-	-	0	56	18	**74**
11	-	-	-	0	8	**8**
12	-	-	-	-	0	**0**

Solution

From Equations 2-8 and 2-10 for $CN = 80$,

$$S = \frac{1000}{80} - 10 = 2.5\ in$$

$$I_a = 0.2S = 0.5\ in$$

The initial abstraction absorbs rainfall up to a value of 0.5 in, including all 0.3 in during the first hour and 0.2 in during the second hour, at which point remaining, continuing losses begin. Cumulative rainfall excess is computed using Equation 2-9. For example, considering the second hour and corresponding cumulative rainfall of 0.8 in,

$$P_e = \frac{(0.8 - 0.5)^2}{0.8 + 0.8(2.5)} = 0.03\ in$$

Computations for remaining hours proceed in a similar manner. Table P2-13 summarizes the computations and lists the resulting distribution of rainfall excess in Column 5.

Table P2-13

(1)	(2)	(3)	(4)	(5)
Time (hr)	Rainfall (in)	Cumulative rainfall (in)	Cumulative rainfall excess, P_e (in)	Rainfall excess (in)
1	0.3	0.3	0.00	**0.00**
2	0.5	0.8	0.03	**0.03**
3	0.7	1.5	0.29	**0.26**
4	0.4	1.9	0.50	**0.21**
5	0.6	2.5	0.89	**0.39**
6	0.5	3.0	1.25	**0.36**
7	0.4	3.4	1.56	**0.31**

Problem 2.14 NRCS curve number method

An urban watershed consists of the following components:

Description	Soil Group	Area (ac)
Residential development (0.5 ac lots; 25% impervious)	C	10
Commercial development (85% impervious)	B	5
Industrial development (72% impervious)	A	5

Moisture conditions for the entire basin are characterized as AMCI (i.e., low moisture). For a storm having a 6-in rainfall, estimate the amount of rainfall excess.

Solution

From Table 2-7, values of *CN* for residential, commercial and industrial components are 80, 92, and 81, respectively. An area-averaged *CN* can be computed as follows:

$$CN_{avg} = \frac{\sum_{i=1}^{n} (CN)_i A_i}{A_{total}} = \frac{[(80 \times 10) + (92 \times 5) + (81 \times 5)]}{20} = 83$$

For given antecedent moisture conditions, the *CN* value should be adjusted using Equation 2-11.

$$CN(I) = \frac{4.2(83)}{10 - 0.058(83)} = 67$$

Excess rainfall is computed using Equations 2-9 and 2-10.

$$S = \frac{1000}{67} - 10 = 4.93 \ in$$

$$P_e = \frac{[6 - 0.2(4.93)]^2}{6 + 0.8(4.93)} = \underline{\underline{2.53 \ in}}$$

Problem 2.15 Natural unit hydrograph

Given the excess rainfall distribution and direct runoff hydrograph below, derive the 1-hr unit hydrograph for the corresponding watershed.

Time (hr)	1	2	3	4	5	6	7	8	9	10	11
Rainfall excess (in)	0.2	0.7	1.4	0.2	-	-	-	-	-	-	-
Direct runoff (cfs)	20	134	454	857	1003	815	554	345	186	74	8

Solution

The number of rainfall excess intervals, *M*, is equal to four, and the number of direct runoff ordinates, *N*, is eleven. Therefore, there will be eight unit hydrograph ordinates (i.e., $N - M + 1 = 8$). From Equation 2-13, a total of eleven equations can be written in terms of eight unknowns, as follows:

$$Q_1 = P_1 U_1$$
$$Q_2 = P_2 U_1 + P_1 U_2$$
$$\ldots$$
$$Q_4 = P_4 U_1 + P_3 U_2 + P_2 U_3 + P_1 U_4$$
$$Q_5 = P_4 U_2 + P_3 U_3 + P_2 U_4 + P_1 U_5$$
$$\ldots$$
$$Q_{10} = P_4 U_7 + P_3 U_8$$
$$Q_{11} = P_4 U_8$$

Only the first eight equations are needed to solve for the unknown unit hydrograph ordinates. As an example, consider $n = 1$ and 2,

$$U_1 = \frac{Q_1}{P_1} = \frac{20}{0.2} = 100\ cfs/in$$

$$U_2 = \frac{Q_2 - P_2 U_1}{P_1} = \frac{134 - 0.7(100)}{0.2} = 320\ cfs/in$$

Remaining computations proceed in a similar manner. Table P2-15 provides a summary of computations and lists the unit hydrograph in Column 3.

Problem 2.16 NRCS dimensional unit hydrograph

Derive the NRCS triangular and curvilinear unit hydrographs for an 8-mi^2 watershed having an average slope of 0.025 ft/ft. The hydraulic flow length from the catchment boundary to the outlet is 2.5 mi (13,200 ft), and the basin is characterized by a curve number of 85.

Solution

Time of concentration, t_c, is computed using the NRCS lag equation given in Table 2-3.

$$t_c = \frac{100(13200)^{0.8}\left[\left(\frac{1000}{85}\right) - 9\right]^{0.7}}{19000(0.025)^{1/2}} = 134.2\ min = 2.2\ hrs$$

Time to peak discharge, t_p, unit hydrograph base time, t_b, peak discharge, Q_p, and effective duration, t_r, are computed as follows:

$$t_p = \frac{2}{3}(2.2) = 1.5\ hrs$$

$$t_b = 1.67(t_p) + t_p = 2.67(1.5) = 4.0 \; hrs$$

$$Q_p = \frac{484(8)}{1.5} = 2580 \; cfs/in$$

$$t_r = 0.133(2.2) = 0.3 \; hrs$$

Table P2-15

(1)	(2)	(3)
Time (hr)	Direct runoff (cfs)	Unit hydrograph (cfs/in)
0	0	**0**
1	20	**100**
2	134	**320**
3	454	**450**
4	857	**370**
5	1003	**250**
6	815	**160**
7	554	**90**
8	345	**40**
9	186	**0**
10	74	-
11	8	-
12	0	-

Thus, the 0.3-hr triangular unit hydrograph can be derived by plotting points (0,0), (1.5, 2580) and (4, 0). The corresponding curvilinear unit hydrograph is found by multiplying values in Table 2-8 by respective values of t_p and Q_p. The two resulting unit hydrographs are shown in Figure P2-15.

Problem 2.17 S-hydrograph method

Convert the 1-hr unit hydrograph provided in Problem 2.11 to a 3-hr unit hydrograph using the S-hydrograph method.

Solution

Table P2-17 summarizes the stepwise computations of the new unit hydrograph (UH). Note that the current duration, t_r, is 1 hr, while the desired duration, t_r', is 3 hrs. Specific entries in the Table are as follows:

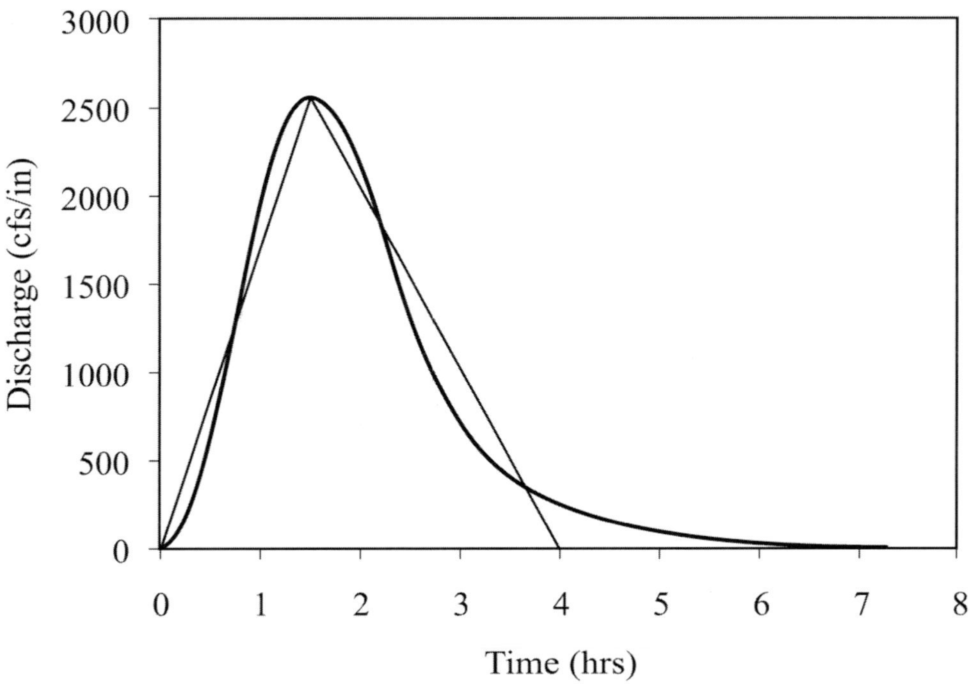

Figure P2-16

Columns (1) and (2): The current 1-hr unit hydrograph
Column (3): Represents a series of unit hydrographs, each lagged by the current duration
Column (4): The 1-hr S-curve is obtained by summing the values in Columns 2 and 3
Column (5): The S-curve from Column 4 lagged by the desired duration of 3 hrs
Column (6): Difference between the current and lagged S-curves
Column (7): The 3-hr unit hydrograph is computed by multiplying the values in Column 6 by the ratio of t_r to t_r', or 0.33.

Table P2-17

(1)	(2)	(3)			(4)	(5)	(6)	(7)
Time (hr)	1-hr UH (cfs/in)	Lagged 1-hr UH (cfs/in)			1-hr S-curve (cfs/in)	Lagged S-curve (cfs/in)	Difference (cfs/in)	3-hr UH (cfs/in)
0	0	-	-	-	0	-	0	**0**
1	100	0	-	-	100	-	100	**33**
2	320	100	0	-	420	-	420	**140**
3	450	320	100	0	870	0	870	**290**
4	370	450	320	100	1240	100	1140	**380**
5	250	370	450	320	1490	420	1070	**357**
6	160	250	370	450	1650	870	780	**260**
7	90	160	250	370	1740	1240	500	**167**
8	40	90	160	250	1780	1490	290	**97**
9	-	40	90	160	1780	1650	130	**43**
10	-	-	40	90	1780	1740	40	**13**
11	-	-	-	40	1780	1780	0	**0**

Problem 2.18 Snyder's synthetic unit hydrograph

Using Snyder's method, derive a 1-hr synthetic unit hydrograph for the basin described in Problem 2.15. Assume that L_{CA} = 1 mi, C_t = 1.9, and C_p = 0.6.

Solution

Lag time is computed using Equation 2-17.

$$t_L = 1.0(1.9)[(2.5)(1.0)]^{0.3} = 2.5\ hrs$$

From Equation 2-18, the standard duration of rainfall excess is

$$t_r = \frac{2.5}{5.5} = 0.45\ hrs$$

However, the desired duration, t_{ra}, is 1 hr, so the lag time should be adjusted according to Equation 2-19.

$$t_{La} = 2.5 + 0.25(1.0 - 0.45) = 2.64\ hrs$$

Equations 2-20 and 2-21 can be used to determine the peak discharge, Q_p, and time to peak discharge, t_p.

$$t_p = 2.64 + 0.5(1.0) = 3.14\ hrs$$

$$Q_p = \frac{640(0.6)(8)}{2.64} = 1164\ cfs/in$$

The unit hydrograph widths at discharges equal to 50 and 75 percent of the peak discharge, from Equations 2-22 and 2-23, are

$$W_{50} = 770\left(\frac{8}{1164}\right)^{1.08} = 3.55\ hrs$$

$$W_{75} = 440\left(\frac{8}{1164}\right)^{1.08} = 2.03\ hrs$$

The unit hydrograph base time, t_b, is computed by finding that which guarantees the area under the curve corresponds to 1 in of rainfall excess. For A in mi^2, Q_p in cfs, and W_{50} and W_{75} in hrs,

$$1\ in = \left[\left(\frac{t_b + W_{50}}{2}\right)\left(\frac{Q_p}{2}\right) + \left(\frac{W_{75} + W_{50}}{2}\right)\left(\frac{Q_p}{4}\right) + \left(\frac{W_{75}}{2}\right)\left(\frac{Q_p}{4}\right)\right](hr\text{-}cfs)\times \left[\left(\frac{1}{A}\right)\left(\frac{1\ mi^2}{(5280)^2\ ft^2}\right)\left(\frac{12\ in}{1\ ft}\right)\left(\frac{3600\ sec}{1\ hr}\right)\right]$$

Solving for t_b yields

$$t_b = 2581\frac{A}{Q_p} - 1.5W_{50} - W_{75} = 2581\left(\frac{8}{1164}\right) - 1.5(3.55) - 2.03 = 10.4\ hrs$$

A total of seven coordinates are now known and can be used to define the resulting unit hydrograph shown in Figure P2-18. From the starting point of (0,0) (i.e., A), the remaining points are as follows:

$$B \rightarrow \left(t_p - \frac{W_{50}}{3}, 0.5Q_p\right) = (1.96, 582)$$

$$E \rightarrow \left(t_p + \frac{2}{3}W_{75}, 0.75Q_p\right) = (4.49, 873)$$

$$C \rightarrow \left(t_p - \frac{W_{75}}{3}, 0.75Q_p\right) = (2.46, 873)$$

$$F \rightarrow \left(t_p + \frac{2}{3}W_{50}, 0.5Q_p\right) = (5.51, 582)$$

$$D \rightarrow (t_p, Q_p) = (3.14, 1164)$$

$$G \rightarrow (t_b, 0) = (10.4, 0)$$

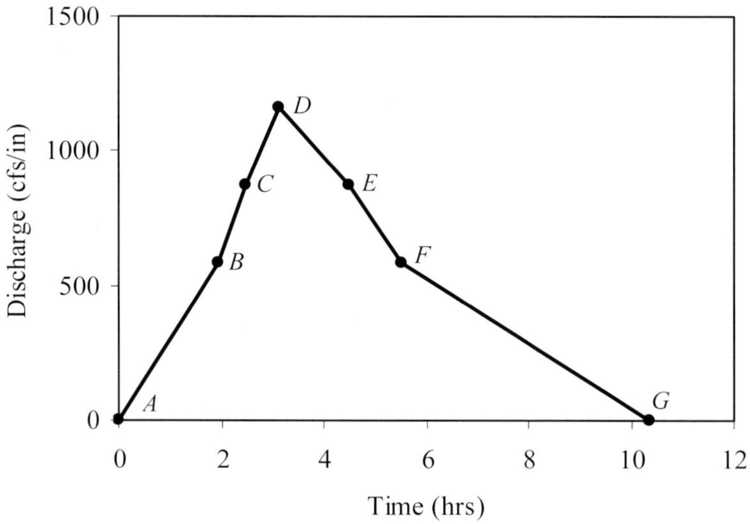

Figure P2-18

Problem 2.19 Clark's synthetic unit hydrograph

Use Clark's method to develop a 1-hr synthetic unit hydrograph for the watershed described in Problem 2.15. Use Equation 2-24 to obtain the time-area diagram. Assume that time of concentration of the watershed is 3 hrs. Use a computational time interval (Δt) of 0.5 hrs, and assume the storage coefficient is $0.6t_c$.

Solution

Table P2-19 summarizes the solution procedure. Entries in each column are as follows:

Column (1): Time from the beginning of effective rainfall at intervals of Δt.
Column (2): Cumulative watershed area, in acres, contributing flows to the watershed outlet at the time. These values are obtained using Equation 2-24. For example at hour one,

$$A_t = 8 \times 640 \times 1.414 \left(\frac{1}{3} \right)^{1.5} = 1393.28 \text{ acres}$$

Column (3): Area of the watershed, in acres, that started contributing flow to the outlet within the time interval. This area is plotted against time (*Column 1*) to produce the time area histogram given in Figure P2-19a.

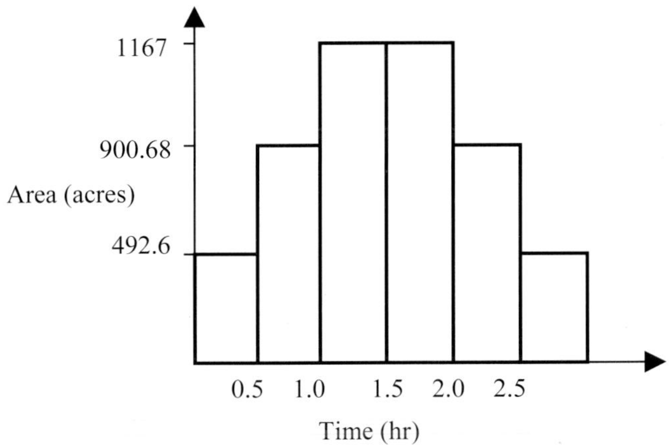

Figure P2-19a

Column (4): Inflow, in cfs, generated by multiplying *Column 3* by 1 inch of rainfall excess, and dividing the resulting value by computational time interval of 0.5 hrs. Example, inflow at 2.5 hrs = 900.68 acres x 1 inch/0.5 hrs. Note that 1acres-inch/hour = 1 cfs.
Column (5): Outflow ordinates obtained by routing the inflow hydrograph through a linear reservoir using Equation 2-27. The storage coefficient, *R*, is commonly considered as the lag time of the watershed (i.e., $0.6t_c$ = 1.8 hrs).

$$C_1 = \frac{0.5}{1.8 + 0.5 \times 0.5} = 0.244$$

$$C_2 = 1 - 0.244 = 0.756$$

Therefore, $Q_t = 0.244 I_t + 0.756 Q_{t-1}$. As an example, outflow at time 2-hrs

$$= 0.244 \times 2334.22 + 0.756 \times 1038.5 = 1354.52\ cfs$$

Table P2-19

(1)	(2)	(3)	(4)	(5)	(6)	(7)
Time (hr)	A_T (acres)	ΔA (acres)	I_t (cfs)	Q_t (cfs)	Q_{IUH} (cfs)	$Q_{1\text{-}hr}$ (cfs)
0.0	0	0	0	0.00	0.00	**0.00**
0.5	492.60	492.60	985.20	240.29	120.14	**60.07**
1.0	1393.28	900.68	1801.36	621.03	430.66	**215.33**
1.5	2559.61	1166.34	2332.67	1038.50	829.77	**474.96**
2.0	3726.72	1167.11	2334.22	1354.52	1196.51	**813.59**
2.5	4627.40	900.68	1801.36	1463.50	1409.01	**1119.39**
3.0	5120.00	492.60	985.20	1346.84	1405.17	**1300.84**
3.5	0	0	0	1018.34	1182.59	**1295.80**
4.0	0	0	0	769.97	894.15	**1149.66**
4.5	0	0	0	582.17	676.07	**929.33**
5.0	0	0	0	440.18	511.17	**702.66**
5.5	0	0	0	332.82	386.50	**531.28**
6.0	0	0	0	251.64	292.23	**401.70**
6.5	0	0	0	190.27	220.95	**303.73**
7.0	0	0	0	143.86	167.06	**229.65**
7.5	0	0	0	108.77	126.32	**173.63**
8.0	0	0	0	82.24	95.51	**131.28**
8.5	0	0	0	62.18	72.21	**99.26**
9.0	0	0	0	47.02	54.60	**75.05**
9.5	0	0	0	35.55	41.28	**56.75**
10.0	0	0	0	26.88	31.21	**42.91**

The solution process is recursive. As a result, outflow ordinates will theoretically continue for an infinite duration. However, it is customary to truncate the recession limb of the hydrograph where the outflow volume exceeds 0.995 inches. Only the first 20 outflow ordinates are given in Table P2-19.

Column (6): Clark's instantaneous unit hydrograph ordinates obtained by averaging *Column 5* values over the computational time step using Equation 2-30.

Column (7): Ordinates of a 1-hr synthetic unit hydrograph (Figure 2-19b) are obtained by lagging the instantaneous unit hydrograph ordinates by an hour, and taking averages of the ordinates of the original and the lagged hydrographs at the time. For example, the ordinate of the 1-hr synthetic unit hydrograph at hour 3 is obtained by averaging ordinates of the instantaneous unit hydrograph at hour 3 (i.e., original Clark's IUH) and at hour 2 (i.e., Clark's IUH lagged by one hour).

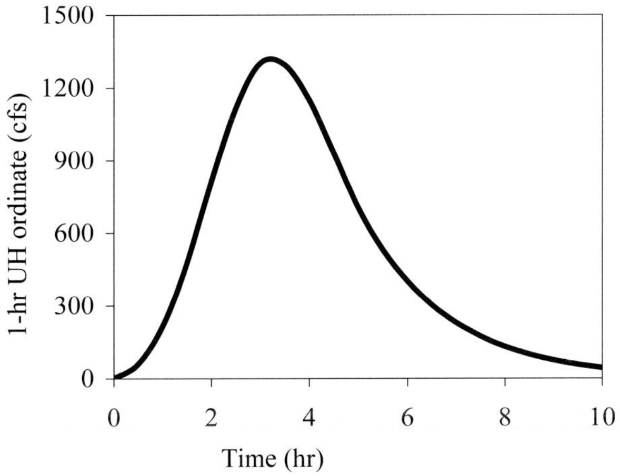

Figure P2-19b

Problem 2.20 Colorado Urban Hydrograph Procedure

Use the Colorado Urban Hydrograph Procedure to derive a 1-hr synthetic unit hydrograph for a 3-mi^2 watershed having an average slope of 0.025 ft/ft. Assume that L_{CA} = 1 mi, L = 2 mi, and that 30 percent of the watershed area is impervious.

Solution

For a watershed that has 30 percent impervious area, the time to peak coefficient C_t (Equation 2-34) and the peaking parameter P (Equation 2-35) are calculated, using the coefficients given in Table 2-9, as

$$C_t = 0.000023 \times 30^2 - 0.00224 \times 30 + 0.146 = 0.0995$$

$$P = 0.00245 \times 30^2 - 0.012 \times 30 + 2.16 = 4.005$$

The unit hydrograph peaking coefficient, C_p (Equation 2-33), is determined as

$$C_p = 4.005 \times 0.0995 \times 3 = 1.195$$

From Equation 2-31, lag time of the watershed is

$$t_L = 0.0995 \times \left(\frac{2 \times 1}{\sqrt{0.025}} \right)^{0.48} = 0.3364 \, hrs.$$

The time to peak, t_p, is

$$t_p = 0.3364 + 0.5 \times 1 = 0.8364 \, hrs$$

From Equation 2-32, the peak flow rate is

$$Q_p = \frac{640 \times 1.195 \times 3}{0.8364} = 2743.185 \, cfs$$

The widths of the unit hydrograph at 50% (Equation 2-36) and 75% (Equation 2-37) of the peak are

$$W_{50} = \frac{500}{\left(\frac{2743.185}{3} \right)} = 0.547 \, hrs$$

$$W_{75} = \frac{260}{\left(\frac{2743.185}{3} \right)} = 0.284 \, hrs$$

Next, W_{50} and the W_{75} are distributed around the peak. The width to the left side of the peak at 50 percent of the peak is the smaller of $0.35W_{50}$ (i.e., 0.191 hrs) and $0.6t_p$ (0.502 hrs), which is 0.191 hrs for this specific problem. The width to the right side of the peak at 50 percent of the peak is $0.65W_{50}$ (i.e., 0.355 hrs). At 75 percent of the peak, width to the left side of the peak equals $0.45W_{75}$, which is 0.1278 hrs, and width to the right side of the peak is $0.55W_{75}$ (i.e., 0.1562 hrs).

Finally, time base of the unit hydrograph is determined so that the area under the curve corresponds to 1-in of rainfall excess. As in Problem 2-17, for A in mi^2, Q_p in cfs, and W_{50} and W_{75} in hrs, solving for t_b yields,

$$t_b = 2581\frac{A}{Q_p} - 1.5W_{50} - W_{75}$$

$$= 2581\left(\frac{3}{2743.2}\right) - 1.5(0.547) - 0.284 = 1.72 \text{ hrs}$$

Figure P2-20 displays the resulting unit hydrograph.

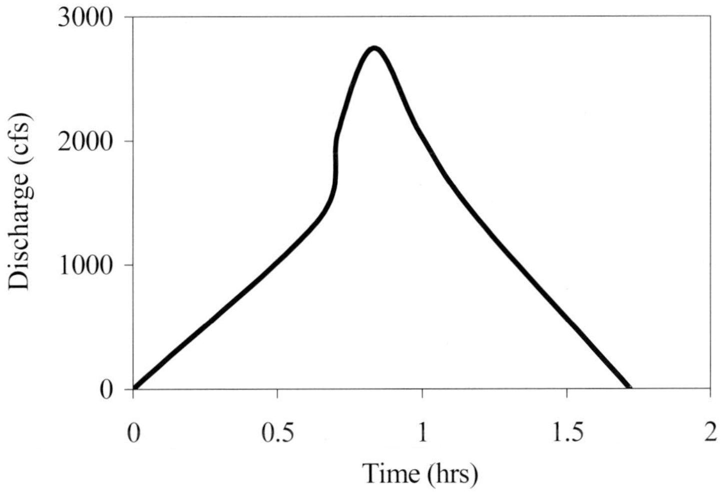

Figure P2-20

Problem 2.21 Tri-triangular Unit Hydrograph

Use the tri-triangular unit hydrograph method to derive a 1-hr synthetic unit hydrograph for an 8-mi^2 watershed. Assume that R_1 = 30%, R_2 = 50%, T_1 = 1 hr, T_2 = 4 hrs, T_3 = 12 hrs, K_1 = 2, K_2 = 3, and K_3 = 3.

Solution

Assuming that rainfall excess of 1-in depth is collected over the 1-hr duration, the total volume of runoff that is generated from the watershed as the result of the rainfall excess would be

$$Area(ft^2) \times Depth(in) = \left(\frac{1}{12}\right)(8 \times 27878400) = 18{,}585{,}600\, ft^3$$

The volume of direct runoff allocated to the first triangle (i.e., the triangle representing fast responding components of the watershed) is

$$R_1 \times 18{,}585{,}600 = 0.3 \times 18{,}585{,}600 = 5{,}575{,}680\, ft^3$$

Likewise, the volume of direct runoff allocated to the second triangle is

$$R_2 \times 18{,}585{,}600 = 0.5 \times 18{,}585{,}600 = 9{,}292{,}800\, ft^3$$

Implying that the remainder of the direct runoff volume (i.e., 18,585,600 - 5,575,680 - 9,292,800 = 3,717,120 ft^3) comes from the third triangle (i.e., the one representing slow responding components of the watershed).

Time bases for triangle 1 (i.e., T_{b1}), triangle 2 (i.e., T_{b2}), and triangle 3 (i.e., T_{b3}) are determined as

$$T_{b1} = T_1 + T_1 K_1 = 1 + 2 \times 1 = 3\ hrs$$

$$T_{b2} = T_2 + T_2 K_2 = 4 + 3 \times 4 = 16\ hrs$$

$$T_{b3} = T_3 + T_3 K_3 = 12 + 3 \times 12 = 48\ hrs$$

Once the total volumes of direct runoff allocated to each triangle and the time base of each triangle is known, peak flow for the triangles (i.e., Q_{p1}, Q_{p2}, Q_{p3}) are calculated as

$$Q_{p1} = \frac{2 \times Volume_1(ft^3)}{T_{b1}(sec)} = \frac{2 \times 5{,}575{,}680}{3600 \times 3} = 1{,}032.5\ cfs$$

$$Q_{p2} = \frac{2 \times Volume_2(ft^3)}{T_{b2}(sec)} = \frac{2 \times 9{,}292{,}800}{3600 \times 16} = 322.67\ cfs$$

$$Q_{p3} = \frac{2 \times Volume_3(ft^3)}{T_{b3}(sec)} = \frac{2 \times 3{,}717{,}120}{3600 \times 48} = 43.02\ cfs$$

Having the time to peaks, the time bases, and the peak flow vales of each triangle, the required 1-hr unit hydrograph could be generated by aggregating flow ordinates of the three triangles at any desired time *t*. Figure P2-21 shows the derived 1 hr synthetic unit hydrograph.

Problem 2.22 Hydrograph routing

Route the direct runoff hydrograph given in Problem 2.14 through the system shown in Figure P2-21a. Specifically, determine the outfall hydrograph from junction *C*. Assume that the system is comprised of point junctions (i.e., no storage capability) and that the average sewer flow time in each length of pipe is 20 min.

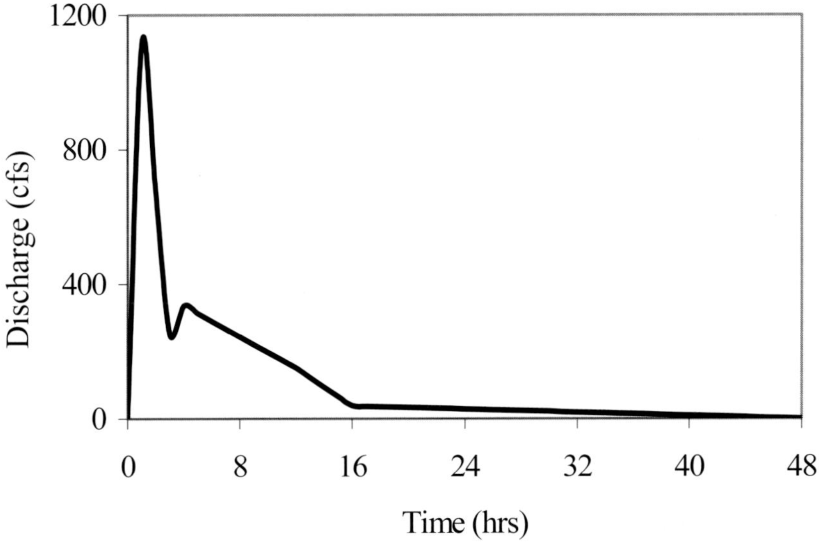

Figure P2-21

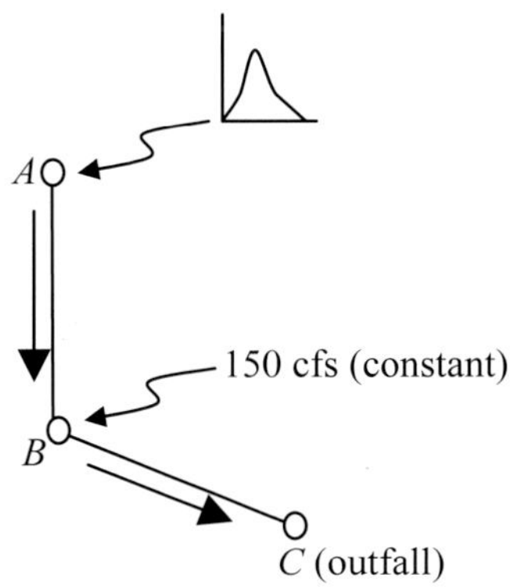

Figure P2-22a

Solution

A simple, but effective, method for routing hydrographs through sewers is to lag the inflow hydrograph by an amount equal to the sewer flow time. The inflow to junction B will thus consist of a duplicate of the original direct runoff hydrograph that is lagged by 20 min. The outflow from B will be the sum of the lagged hydrograph and the constant 150 cfs of additional runoff. Finally, the outfall hydrograph will be the outflow from junction *B* lagged by another 20 min. Figure P2-22b illustrates the *outflow* hydrograph from each of the three junctions.

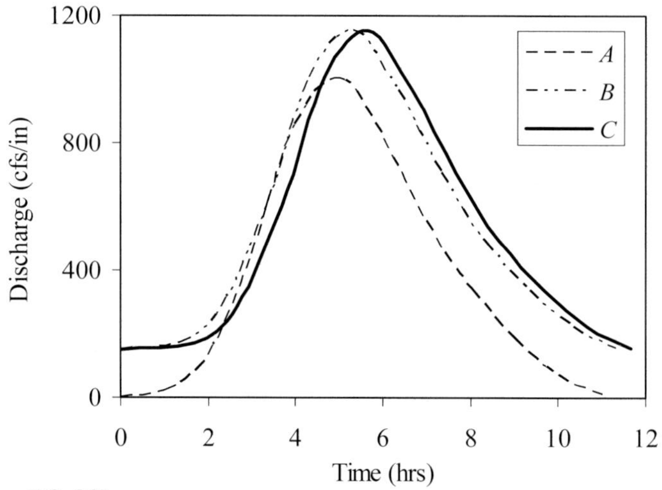

Figure P2-22b

Comments: This particular routing technique is a lumped method that does not consider the unsteady and nonuniform nature of sewer flow. It approximately accounts for sewer flow time, but offers no simulation of wave attenuation. However, interpolation within the computational procedure introduces some numerical attenuation. In addition, if storage junctions are included in the system, consideration must be given to the rate of accumulation or depletion of fluid (i.e., *dS/dt*) at those locations.

REFERENCES CITED

Abbott, M.B., "An Introduction to the European Hydrological System - Système Hydrologique Européen, 'SHE' – 2: Structure of a Physically-Based, Distributed Modeling System," *J. of Hydrology*, vol. 87, 61-77, 1986.

American Society of Civil Engineers (ASCE), *Gravity Sanitary Sewer Design and Construction*, ASCE Manuals and Reports on Engineering Practice No. 60 and Water Pollut. Control Fed. Manual of Practice FD-5, New York, NY, 1982.

American Society of Civil Engineers (ASCE), *Design and Construction of Urban Stormwater Management Systems*, ASCE Manuals and Reports on Engineering Practice No. 77 and Water Pollut. Control Fed. Manual of Practice RD-20, New York, NY, 1992.

Aron, G., and C.E. Egborge, "A Practical Feasibility Study of Flood Peak Abatement in Urban Areas," Report, U.S. Army Corps of Engineers, Sacramento District, Sacramento, CA, 1973.

Babbitt, H.E. and E.R. Baumann, *Sewerage and Sewage Treatment*, John Wiley & Sons Inc., New York, NY, 1958.

Borah, D.K., "Dynamic Modeling and Monitoring of Water, Sediment, Nutrients, and Pesticides in Agricultural Watersheds During Storm Events," Contract Rep. 655, Illinois State Water Survey, 1999.

Boulos, P.F., *Users Guide for H_2OMAP Sewer Pro.* MWH Soft, Inc., 300 North Lake Avenue, Suite 1200, Pasadena, CA, 2004a.

Boulos, P.F., *Users Guide for InfoSewer Pro.* MWH Soft, Inc., 300 North Lake Avenue, Suite 1200, Pasadena, CA, 2004b.

Chen, C., "Rainfall Intensity-Duration-Frequency Formulas," *J. of Hydraulic Engrg.*, ASCE, vol. 109, no. 12, 1603-1621, 1983.

Chow, V.T., D.R. Maidment, and L.W. Mays, *Applied Hydrology*, McGraw-Hill, New York, NY, 1988.

Clark, C. O., "Storage and the Unit Hydrograph," *Transaction of the Amer. Soc. Civ. Engrg.*, vol 110, 1945.

Engman, E.T., "Roughness Coefficients for Routing Surface Runoff," *J. of Irrigation and Drainage Engrg.*, ASCE, vol. 112, no. 1, 39-53, 1986.

Ewen, J., G. Parkin, and P.E. O'Connell, "SHETRAN: Distributed River Basin Flow and Transport Modeling System," *J. of Hydrologic Engrg.*, ASCE, vol. 5, no. 3, 250-258, 2000.

Fair, G.M., J.C. Geyer, and D.A. Okun, *Elements of Water Supply and Wastewater Disposal*, Wiley, New York, NY, 1971.

Federal Aviation Administration (FAA), Circular on Airport Drainage, Report A/C 050-5320-5B, U.S. Department of Transportation, Washington, DC, 1970.

Hershfield, D.M. "Rainfall Frequency atlas of the United States for Durations from 30 Minutes to 24 Hours and Return Periods from 1 to 100 Years," Technical Paper 40, U.S. Department of Commerce, Weather Bureau, Washington, DC, 1961.

Huber, W.C., and R.E. Dickinson, *Storm Water Management Model, Version 4, User's Manual*, EPA/600/3-88/001a, U.S. Environmental Protection Agency, Athens, GA, 1988.

Hydrologic Engineering Center (HEC), *HEC-1 Flood Hydrograph Package - User's Manual*, U.S. Army Corps of Engineers, Davis, CA, 1990.

Hydrologic Engineering Center (HEC), *Hydrologic Modeling System HEC-HMS - User's Manual*, U.S. Army Corps of Engineers, Davis, CA, 2000.

Izzard, C.F., "Hydraulics of Runoff from Developed Surfaces," *Proc. Highway Research Board*, vol. 26, 129-146, 1946.

Julien, P.Y., and B. Saghafian, "CASC2D User's Manual," Dept. of Civil Engrg. Rep., Colorado State Univ., Fort Collins, CO, 1991.

Kirpich, Z.P., "Time of Concentration of Small Agricultural Watersheds," *Civ. Engrg.*, vol. 10, no. 6, 362, 1940.

Leavesley, G.H., R.W. Lichty, B.M. Troutman, and L.G. Saindon, "Precipitation-Runoff Modeling System – User's Manual," U.S. Geological Survey, Water-Resources Investigations Report 83-4238, 1983.

Mays, L.W., *Water Resources Engineering*, Wiley, New York, NY, 2001.

Metcalf and Eddy, Inc., *Wastewater Engineering: Treatment, Disposal, and Reuse*, 3rd ed., McGraw-Hill, New York, NY, 1991.

Morgali, J.R., and R.K. Linsley, "Computer Analysis of Overland Flow," *J. Hydraulic Div.*, ASCE, vol. 91, no. HY3, 81-100, 1965.

Ogden, F.L., "CASC2D Version 1.18 Reference Manual," Dept of Civil and Environ. Engrg., Rep. U-37, CT1665-1679, Univ. of Connecticut, Storrs, CT, 1998.

Refsgaard, J.C., and B. Storm, "Chapter 23: MIKE SHE," in *Computer Models of Watershed Hydrology*, ed. by V.P. Singh, Water Res. Pub., Littleton, CO, 1995.

Rossman, L.A., "*Storm Water Management Model Users Manual Version 5*", U.S. Environmental Protection Agency, National Risk Management Research Laboratory, Cincinnati, OH, 2004.

Singh, V.P., and D.A. Woolhiser, "Mathematical Modeling of Watershed Hydrology," *J of Hydrologic Engrg.*, ASCE, vol. 7, no. 4, 270-292, 2002.

Smith, R.E., Discussion of "Runoff Curve Number: Has it Reached Maturity?" by V.M. Ponce and R.H. Hawkins. *J. of Hydrologic Engrg.*, ASCE, vol. 2, no. 3, 145-147, 1997.

Snyder, F.F., "Synthetic Unit Hydrographs," *Trans. Amer. Geophysical Union*, vol. 19, 447-454, 1938.

Soil Conservation Service (SCS), "Computer Model for Project Formulation Hydrology," Technical Release No. 20, U.S. Dept. of Agriculture, Washington, DC, 1965.

Soil Conservation Service (SCS), *National Engineering Handbook*, U.S. Dept. of Agriculture, Washington, DC, 1985.

Soil Conservation Service (SCS), "Urban Hydrology for Small Watersheds," Technical Release 55, U.S. Dept. of Agriculture, Washington, DC, 1986.

Urban Drainage and Flood Control District (UDFCD), "Colorado Urban Hydrograph Procedure," *Users Manual*, Denver, CO, 2001.

UDFCD, "Urban Storm Drainage Criteria Manual Rev. Ed.,"*Users Manual*, Denver Regional Council of Governments, Denver, CO, 1984.

U.S. Public Health Service (USPHS), "Manual of Septic-Tank Practice," PHS Publication No. 526, U.S. Dept. of Health, Washington, DC, 1963.

Viessman, W., and G.L. Lewis, *Introduction to Hydrology*, 4th ed., Harper Collins, New York, NY, 1996.

Westphal, J.A. "Design of Storm Water Inlets," in *Stormwater Collection Systems Handbook*, ed. by L.W. Mays, McGraw Hill, New York, NY, 2001.

Woolhiser, D.A., R.E. Smith, and D.C. Goodrich, "KINEROS – A Kinematic Runoff and Erosion Model: Documentation and User Manual," Rep. No. ARS-77, U.S. Dept. of Agriculture, Washington, DC, 1990.

Yen, B.C., and V.T. Chow, "Local Design Storms, Vol. I to III," Report No. FHWA-RD-82-063 to 065, U.S. Dept. of Transportation, Fed. Highway Administration, Washington, DC, 1983.

CHAPTER THREE
Hydraulics of Steady Sewer Flows

3.1 *INTRODUCTION*

Steady flows are those in which velocity, flow depth, and other flow properties at a particular location in the system are constant over time. Strictly speaking, sewer flows are unsteady. A majority of sewer analyses, however, are performed using the steady flow assumption over the time required for flow to pass through the sewer. This is a reasonable assumption in most instances since the duration of flows is much greater than flow travel time; that is, flow changes occur slow enough to be regarded as quasi-steady. However, for applications in which the propagation of pressure or surface waves significantly influences overall system behavior, unsteady flow concepts are required (see Chapter 4).

In application of the hydraulic principles described herein, a number of key geometric elements are important for describing flow. Table 3-1 provides a list of convenient formulae for computing common geometric properties for a variety of section shapes. In review of the Table, it is worth noting that a large majority of sewers constructed in the United States in recent history are circular in cross section. Previously, however, a wide variety of irregular sections were common, including egg-shaped, semi-elliptical, horseshoe, oval, and others. Since circular sections have become far more commercially available and are subjected to significantly improved finishing techniques, circular sections have now become a standard cross-sectional shape.

3.2 *EQUATIONS OF MOTION*

The governing equations for one-dimensional flow through sewers are based on conservation of mass, energy, and momentum. They are applicable regardless of whether conditions are classified as open channel flow, in which sewers flow partially full and flow is driven by gravity, or as pressurized flow.

3.2.1 Continuity Principle

Conservation of mass for steady, constant density flow implies that volumetric flow rate, Q, at different sewer sections remains equal, provided there is no lateral inflow or outflow over the length being considered. Referring to Figure 3-1, the continuity equation is expressed as

$$Q_1 = Q_2 \tag{3-1}$$

or

$$V_1 A_1 = V_2 A_2 \tag{3-2}$$

where V is mean flow velocity and A is effective flow area.

Table 3-1: Common geometric properties

Shape	Area (A)	Wetted Perimeter (P)	Hydraulic radius (R)	Top width (B)
Rectangular 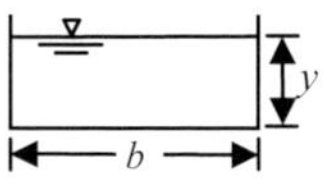	by	$b+2y$	$\frac{by}{b+2y}$	b
Trapezoidal[a] 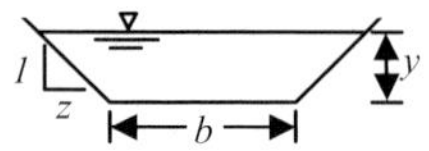	$(b+zy)y$	$b+2y\sqrt{1+z^2}$	$\frac{(b+zy)y}{b+2y\sqrt{1+z^2}}$	$b+2zy$
Parabolic[b] 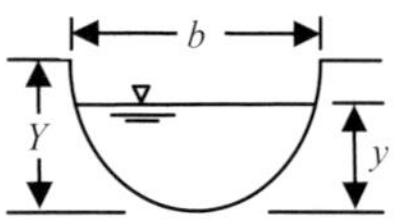	$\frac{2}{3}By$	$\frac{B}{2}\left[\sqrt{1+x^2}+\left(x^{-1}\right)ln\left(x+\sqrt{1+x^2}\right)\right]$	$\frac{4y}{3}\left[\sqrt{1+x^2}+\left(x^{-1}\right)ln\left(x+\sqrt{1+x^2}\right)\right]^{-1}$	$b\sqrt{\frac{y}{Y}}$
Circular[c] 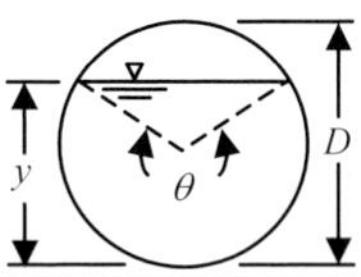	$\frac{(\theta-sin\theta)D^2}{8}$	$\frac{\theta D}{2}$	$\frac{D}{4}\left(1-\frac{sin\theta}{\theta}\right)$	$D\,sin\frac{\theta}{2}$ or $2\sqrt{y(D-y)}$

[a] For triangular sections, use $b = 0$

[b] $x = 4y/B$

[c] use θ in radians, where $\theta = 2cos^{-1}\left(1-\frac{2y}{D}\right)$

Source: Adapted from Chow (1959)

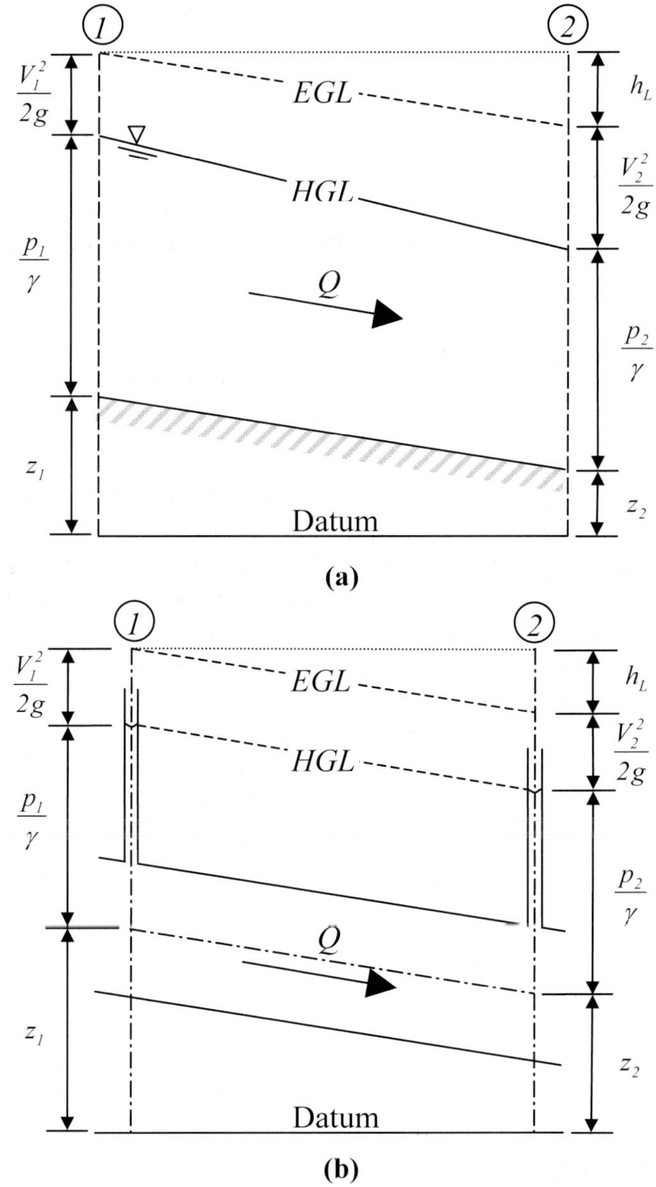

Figure 3-1: Definition sketch for (a) open channel flow, and (b) pressurized pipe flow

3.2.2 Energy Principle

Conservation of energy involves a balance in total energy, expressed as head, between any upstream point of flow and a corresponding downstream point, including head losses caused by friction and the viscous dissipation of turbulence at bends and other appurtenances (i.e., form losses or minor losses). For a uniform velocity distribution, which is synonymous with a fully turbulent flow regime and is generally common for sewer flows, the energy equation can be written as

$$\frac{p_1}{\gamma} + z_1 + \frac{V_1^2}{2g} + h_p = \frac{p_2}{\gamma} + z_2 + \frac{V_2^2}{2g} + h_L \quad (3\text{-}3)$$

where p is fluid pressure; γ is specific weight of the fluid; p/γ is referred to as pressure head; z is elevation of the pipe centerline for pressurized systems or the channel invert for gravity systems, both with respect to a specified datum; V is mean flow velocity; g is acceleration due to gravity; $V^2/2g$ is referred to as velocity head; h_p is head supplied by a pump, if one exists; and h_L represents friction and form losses. The sum of pressure head and elevation terms at any one section defines the location of the hydraulic grade line (HGL), while the addition of velocity head to the HGL describes the energy grade line (EGL).

For most gravity sewer flows, the pressure distribution is nearly hydrostatic (i.e., pressure is a linear function of depth). If the datum is taken as the channel or pipe invert, the sum of elevation and pressure terms from Equation 3-3 is equal to the depth of flow, y, and energy at any location can be expressed as

$$E = y + \frac{V^2}{2g} = y + \frac{Q^2}{2gA^2} \quad (3\text{-}4)$$

where E is referred to as specific energy.

3.2.3 Momentum Principle

Conservation of momentum relates applied forces to changes in linear momentum, which are evidenced by changes in direction or magnitude of flow velocity (i.e., nonuniform flow). Relevant forces include those transmitted to the fluid from pipe or channel walls and those originating from the pressure variation through the fluid. For any direction s, the momentum equation for turbulent flows can be written as

$$\sum F_s = \rho Q(V_{2,s} - V_{1,s}) \tag{3-5}$$

where $\sum F_s$ is the resultant of forces in the s direction, including hydrostatic pressure forces (i.e., pressure multiplied by flow area), gravity forces, and other externally applied forces; ρ is the fluid density; and $V_{2,s}$ and $V_{1,s}$ refer to mean flow velocities in the s direction at downstream and upstream cross sections, respectively.

3.3 CRITICAL FLOW

The relationship between specific energy and depth of flow from Equation 3-4 can prove especially useful in analyzing the spatial variation of gravity flows. Figure 3-2 plots the relationship for a constant flowrate and is referred to as a specific energy diagram.

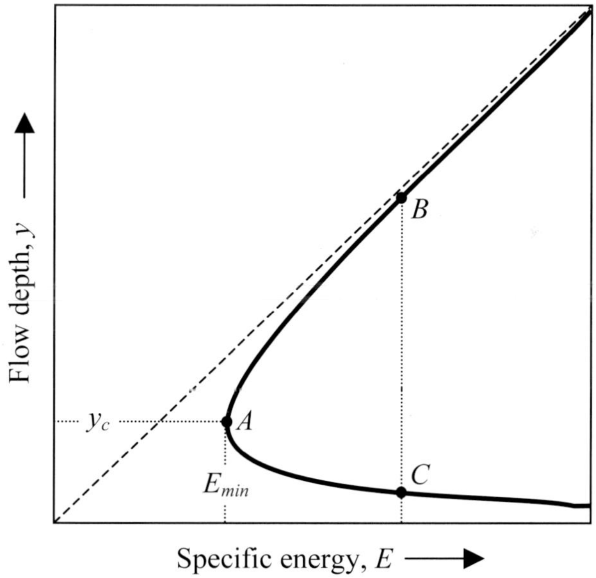

Figure 3-2: Specific energy diagram

Several aspects of the diagram are noteworthy: (1) the lower portion of the curve, AC, asymptotically approaches the abscissa, where $y = 0$; (2) the upper portion of the curve, AB, asymptotically approaches a 45° line (i.e., $E = y$); (3) for any level of specific energy other than the minimum, two alternate depths (i.e., points B and C) exist, and; (4) the curve passes through a minimum energy state at A, where dE/dy is zero and where the depth assumes a critical value, y_c.

Equating the derivative of Equation 3-4 with respect to y to zero and simplifying yields

$$\frac{BQ^2}{gA^3} = 1 \tag{3-6}$$

or, alternatively,

$$\frac{V}{\sqrt{gD}} = 1 \tag{3-7}$$

where D is hydraulic depth, defined as the ratio of area to top width of flow (i.e., A/B).

The left side of Equation 3-6 is the Froude number, Fr, representing a dimensionless ratio of inertial forces to gravity forces. When Fr assumes a value of unity, critical conditions prevail and the flow rate is at a maximum for a given cross sectional and longitudinal slope. Critical flow commonly occurs near free overfalls in mild-sloped channels or near transitions to steep channels and can occur as a result of changes in conduit geometry or invert elevation. Since velocity, flow area, and top width are directly related to flow depth, either Equation 3-6 or 3-7 can be used to solve for y_c for any section shape. For circular pipes of diameter D, critical depth can be approximated as (French, 1985)

$$y_c = \left(\frac{1.01}{D^{0.26}}\right)\left(\frac{Q^2}{g}\right)^{0.25} \tag{3-8}$$

for cases when $0.2D \leq y_c \leq 0.85D$.

When Fr < 1.0, or depths are greater than y_c, conditions are referred to as subcritical, characterized as deep, tranquil flow. Physically, subcritical flow is downstream controlled since downstream disturbances are capable of affecting upstream conditions. Alternatively, when $Fr > 1.0$, or depths are less than y_c, flow is supercritical and is described as being shallow and rapid. In this case, flow is upstream controlled since downstream disturbances cannot propagate upstream faster than the velocity of flow and, thus, will have no effect on upstream conditions. Referring back to Figure 3-2, the upper and lower portions of the specific energy diagram can be interpreted as subcritical and supercritical depths, respectively. The former is more often preferred when dealing with sewer flows, as long as minimum self-cleansing velocities are maintained.

3.4 FLOW RESISTANCE

Whether analyzing flow through pressure or gravity sewers, it is necessary to know relevant characteristics of the conveyance system, such as roughness and configuration, and their ultimate impact on the rate or velocity of flow. The Chezy, Manning, Hazen-Williams, and Darcy-Weisbach formulae are common techniques for evaluating the effects of flow resistance. The Hazen-Williams equation is restricted in application to pressurized sewers, while the remaining techniques are applicable to either force or gravity mains.

Each of the four resistance equations relates velocity, and thus flow rate, to conduit geometry, boundary roughness, and associated energy loss. In application, the relationships can be employed in a variety of ways. For given geometric characteristics of flow, they can be solved to predict flow velocity or discharge. If velocity is known, the formulae can be used to directly evaluate energy lost due to friction. Alternatively, in a typical design problem, they may be solved for the pipe or channel size that can accommodate the estimated peak discharge.

3.4.1 Chezy (Kutter) Equation

The Chezy equation can be expressed as

$$V = C\sqrt{RS_f} \tag{3-9}$$

where C is the Chezy coefficient; R is the hydraulic radius, defined as the ratio of flow area to wetted perimeter, P; and S_f is the friction slope, or slope of the EGL, which is equivalent to energy lost due to friction per unit length of conduit (i.e., h_f/L). The Chezy coefficient can be evaluated empirically by Kutter's equation, expressed as

$$C = \frac{K_1 + \dfrac{K_2}{S_f} + \dfrac{K_3}{n}}{1 + n\left(K_1 + \dfrac{K_2}{S_f}\right)R^{-1/2}} \tag{3-10}$$

where K_1, K_2, and K_3 are constants equal to 41.65, 0.00281, and 1.811, respectively, in U.S. customary units and 23.0, 0.00155, and 1.0 in S.I. units. The value of n in Equation 3-10 is known as Kutter's n, which for all practical purposes is equivalent to the Manning roughness coefficient, also identified as n (see Table 3-2).

Table 3-2: Typical values of resistance coefficients

Conduit material	k_s (Darcy-Weisbach) (ft)	k_s (Darcy-Weisbach) (m)	C_H (Hazen-Williams)	n (Manning)
Cast iron	8.5×10^{-4}	2.5×10^{-4}	130	0.013
Galvanized iron	5×10^{-4}	1.5×10^{-4}	120	0.016
Riveted steel	1.5×10^{-2}	4.5×10^{-3}	110	0.015
Welded steel	2×10^{-4}	6×10^{-5}	120	0.012
Galvanized iron	5×10^{-4}	1.5×10^{-4}	135	0.016
Brass, glass	5×10^{-6}	1.5×10^{-6}	135	0.011
Wood stave	6×10^{-4}	1.8×10^{-4}	120	0.012
Concrete pipe	4×10^{-3}	1.2×10^{-3}	110	0.013
Vitrified clay	6×10^{-3}	1.8×10^{-3}	110	0.013
PVC	5×10^{-6}	1.5×10^{-6}	140	0.009
Asbestos cement	5×10^{-6}	1.5×10^{-6}	140	0.013
Brick	2×10^{-3}	6×10^{-4}	100	0.015
Corrugated metal	1.5×10^{-1}	4.5×10^{-2}	60	0.024
Clay drainage tile	2.5×10^{-3}	7.5×10^{-4}	100	0.013
Smooth earth	-	-	-	0.018
Gravel	-	-	-	0.023
Natural channels	-	-	-	0.025 - 0.08

Source: Adapted from Chow (1959), Lindeburg (1992), WEF (1993), Crowe et al. (2001), ASCE (1982), and others

3.4.2 Manning Formula

In practice, the Manning equation is the most common method for evaluating the effects of flow resistance in gravity sewers. The equation was derived by substituting the following empirical relationship for the Chezy coefficient into Equation 3-9:

$$C = K_m \frac{R^{1/6}}{n} \tag{3-11}$$

where K_m is a constant equal to 1.49 in U.S. customary units and 1.0 in S.I. units. The substitution yields the Manning equation, expressed as

$$V = \frac{K_m}{n} R^{2/3} S_f^{1/2} \tag{3-12}$$

Strictly speaking, the Manning roughness coefficient varies with the geometry of flow and other factors. For fully turbulent flows, however, this variation is negligible. Subsequently, Henderson (1966) showed that a constant roughness, and thus applicability of Equation 3-12, is valid for $n^6 \sqrt{RS_f} \geq 1.9 \times 10^{-13}$, where R is in feet. Furthermore, for sanitary sewer systems, experiments using small-diameter pipes indicate that after a short period of use, n values depend primarily on the biofilm growth on the conduit wall. Therefore, for sanitary sewers having a diameter of less than 3 ft (0.9 m), a Manning roughness value of 0.012 is generally considered satisfactory. For larger pipes, and for stormwater and combined sewer systems, roughness values depend on materials used for construction, as indicated in Table 3-2. Note that n values are provided for guidance only and that adjustments may be warranted based on pipe irregularities or obstructions, alignment, and age or condition of the conveyance system. The most reliable method for obtaining roughness values is to evaluate n using measured depth and discharge records.

3.4.3 Hazen-Williams Formula

The Hazen-Williams equation can be expressed as

$$V = K_H C_H R^{0.63} S_f^{0.54} \tag{3-13}$$

where K_H is a constant equal to 1.318 in U.S. customary units and 0.849 in S.I. units and C_H is the Hazen-Williams roughness coefficient, which can be estimated using Table 3-2. It is important to note that while Equation 3-13, along with associated coefficients, was developed experimentally for potable water systems having a particular range of diameters, flow regimes, and friction slopes, it is often used indiscriminately in sewer design and analysis because of its relative simplicity. Several authors, including Jain et al. (1978) and Bombardelli and Garcia (2003), have shown that using the equation for conditions outside its range of validity can induce significant errors (e.g., ±40 percent). Thus, use of the Hazen-Williams formula should generally be restricted to pressurized systems carrying fluid at temperatures, and thus viscosities, similar to those experienced in potable water systems and to diameters and flow regimes similar to those used in development of the formula. General guides for applicability include temperatures near 60°F (16°C), diameters between 2 in (50 mm) and 6 ft (1.85 m), and velocities less than 10 fps (3 m/s).

3.4.4 Darcy-Weisbach Formula

The Darcy-Weisbach equation is a theoretically-based, but empirical, formula that is most commonly used in the analysis of pressurized sewer systems. It is a variant of the Prony equation (Gaspard de Prony, France), which was developed by Henry Darcy of France and further refined into the form used today by Julies Weisbach of Saxony in 1845. The formula is generally expressed as

$$V = \sqrt{\frac{8gRS_f}{f}} \tag{3-14}$$

where f is the Darcy-Weisbach friction factor, which can be evaluated as a function of equivalent sand grain roughness, k_s, and Reynolds number, Re. Reynolds number is a dimensionless ratio of inertial forces to viscous forces acting on flow and is defined for any cross-sectional shape as

$$Re = \frac{V4R}{\upsilon} \tag{3-15}$$

where ν is the fluid's kinematic viscosity. For $Re < 2{,}000$, flow is referred to as laminar; whereas for $Re > 4{,}000$, flow is generally turbulent, which is typical of most sewer flows. Between 2,000 and 4,000 is a transitional region of flow.

While several formulae have been proposed for the Darcy-Weisbach friction factor, for full flowing commercially available conduits, f is often evaluated by the Colebrook-White equation, expressed as

$$\frac{1}{\sqrt{f}} = -2\log\left(\frac{k_s}{14.83R} + \frac{2.52}{Re\sqrt{f}}\right) \tag{3-16}$$

which is approximately valid for transitional and fully turbulent flows. The relationship between f and Re for open channel flow, however, differs as a result of the interdependence of hydraulic radius, discharge and friction slope (Henderson, 1966). The Colebrook-White equation for gravity systems is written with slightly modified coefficients as follows:

$$\frac{1}{\sqrt{f}} = -2\log\left(\frac{k_s}{12R} + \frac{2.5}{Re\sqrt{f}}\right) \tag{3-17}$$

Notice that Equations 3-16 and 3-17 are implicit since the friction factor appears on both sides of the equation, and thus, they require a trial-and-error solution. Swamee and Jain (1976) developed a more convenient formula that circumvents the iterative solution. The Swamee-Jain equation written explicitly for friction factors in circular pipes having a diameter D is

$$f = \frac{1.325}{\left[ln\left(\frac{k_s}{3.7D} + \frac{5.74}{Re^{0.9}}\right)\right]^2} \quad \text{or} \quad f = \frac{0.25}{\left[log\left(\frac{k_s}{3.7D} + \frac{5.74}{Re^{0.9}}\right)\right]^2} \tag{3-18}$$

The equation is valid for $10^{-6} \leq k_s/D \leq 2 \times 10^{-2}$ and $4 \times 10^3 \leq Re \leq 10^8$, which covers most hydraulic problems involving full flow through commercially available conduits.

3.5 UNIFORM FLOW

If hydraulic characteristics such as discharge and flow area, and therefore velocity, remain the same over the length of a sewer, flow is referred to as uniform. Flow will inevitably approach such conditions if the corresponding channel is sufficiently long. For contrasting nonuniform, or varied, conditions, flow can either be described as rapidly varied, in which changes occur abruptly over short distances, or gradually varied. For example, full flow through a constant diameter pipe would be described as uniform; whereas flow over a weir could be categorized as rapidly varied flow.

For gravity sewers, the concept of uniform flow is more complicated than that for full flowing pressurized pipes since properties such as area, wetted perimeter, and hydraulic radius are not constant. In this case, the uniform flow depth is referred to as the normal depth, y_n. Since the normal depth and velocity remain constant over the length of channel considered, the slope of the HGL (i.e., fluid surface) is parallel to both the friction slope, S_f, and the slope of the pipe or channel, S_o. Under these conditions, gravity forces in the direction of flow are in balance with opposing resistance forces acting at the perimeter of the conduit.

Uniform flow in open channels can be evaluated using the Chezy, Manning or Darcy-Weisbach equations (i.e., Equations 3-9, 3-12, or 3-14) if the channel slope is substituted for S_f. For example,

$$Q = \frac{K_m}{n} AR^{2/3} S_o^{1/2} \tag{3-19}$$

is the Manning equation for computing uniform flow rate. In application, it is often necessary to solve Equation 3-19 for pipe diameter as a function of full-flowing discharge, or

$$D = \left(\frac{K_D Q n}{S_o^{1/2}} \right)^{3/8} \tag{3-20}$$

where K_D is a constant equal to 2.16 in U.S. customary units and 3.21 in S.I. units. A similar expression can be written using the Darcy-Weisbach equation,

$$D = \left(\frac{0.811 f Q^2}{g S_o} \right)^{1/5} \tag{3-21}$$

It is important to emphasize that Equations 3-20 and 3-21 apply to the case when the normal depth equals the pipe diameter and hydraulic radius is equal to $D/4$, or alternatively stated, when the pipe is flowing under just-full, gravity conditions. Thus, for a preselected invert slope and design discharge, based on an analysis of wastewater loads, the rational equation, or a more detailed hydrologic analysis, D represents the minimum pipe diameter required to maintain gravity flow. For practical purposes, D should then be rounded up to the nearest commercially-available pipe diameter. Once the diameter is known, velocity should be checked to ensure self-cleansing capability. If inadequate, a new slope should be selected and the process repeated.

3.6 GRADUALLY VARIED FLOWS

When gravity forces are unbalanced with resisting shear forces, flow is nonuniform, or varied. Consequently, flow will either be accelerated or retarded, and the depth of flow in a gravity sewer will vary longitudinally along its conduit. Such conditions are common at grade and cross section changes, which induce new normal and critical depths, or near hydraulic structures (e.g., weir, sluice gate, or channel transition), where unique relationships exist between depth and discharge, potentially preventing uniform flow. If resulting changes in depth, and thus velocity, occur gradually over long distances, the effects of acceleration on flow between two adjacent sections is negligible. Flow can thus be analyzed directly using energy principles, and conditions are classified as being gradually varied.

3.6.1 Gradually Varied Flow Equation

In addition to the basic assumption, application of gradually varied flow principles assumes that friction loss is identical to that for uniform flow having the same velocity and hydraulic radius, and thus can be computed from the Chezy, Manning, or Darcy-Weisbach equations; that S_o is small (i.e., < 10 percent) so that the pressure distributions remain hydrostatic; that velocity distributions are constant; that resistance coefficients are constant along the length of channel being considered and are independent of flow depth; and that the channel has a constant cross-section shape and slope (i.e., prismatic). Under these conditions, the energy equation can be used to derive the following two alternate forms of the gradually varied flow equation:

$$\frac{dE}{dx} = S_o - S_f \tag{3-22}$$

and

$$\frac{dy}{dx} = \frac{S_o - S_f}{1 - Fr^2} \tag{3-23}$$

where x is the longitudinal length, from upstream to downstream, of the channel being considered and where S_f is often taken as the arithmetic mean slope of the EGL between two adjacent cross sections of interest. These relationships are useful in both qualitative and quantitative aspects for evaluating water surface profiles.

3.6.2 Classification of Water Surface Profiles

Equation 3-23 can be used to qualitatively evaluate gradually varied flow profiles on various channel slopes. The evaluation and subsequent classification of shapes is an important preliminary step to computing associated water surface profiles since it yields information about the location of hydraulic controls, the direction of computation (i.e., upstream for subcritical flow or downstream for supercritical flow), and whether depth is increasing or decreasing. Classification of profiles begins by establishing criteria upon which the water surface depends, namely slope of the channel and depth of flow relative to the critical and normal depths (i.e., y_c and y_n). The criteria are as follows:

- When $y_n > y_c$, the slope is mild and is designated as M
- When $y_n < y_c$, the slope is steep and is designated as S

- When $y_n = y_c$, the slope is critical and is designated as *C*
- If lines are drawn to represent the normal and critical depths (i.e., NDL and CDL), three regions of flow are created; Zone 1 is the region of largest depth, Zone 2 lies between the NDL and CDL, and Zone 3 lies adjacent to the channel bottom.

These criteria can be used to define the water surface profiles shown in Figure 3-3. For example, an *M1* profile exists when $y > y_n > y_c$. Under these conditions, $Fr < 1$ since $y > y_c$, and $S_f < S_o$ since $y > y_n$. Thus, the sign of the numerator and denominator in Equation 3-23 are both positive, and dy/dx must be positive. The implication is that depth will increase with x, typically referred to as a backwater curve. Flow will also approach the normal depth upstream. Alternatively, for an *M2* profile, $y_n > y > y_c$, so that the sign of dy/dx must be negative. In this case, depth decreases with x, which represents a drawdown curve. Similar analyses can be made for other profiles. Note that horizontal and adverse slopes are unique cases in that the normal depth does not exist for either case. In addition, certain theoretical aspects of some profiles do not correlate well with realistic, physical behavior. For example, complete development of an *M3* profile is unlikely; upstream flow cannot be zero, and the downstream end of the profile will likely be overcome by a hydraulic jump.

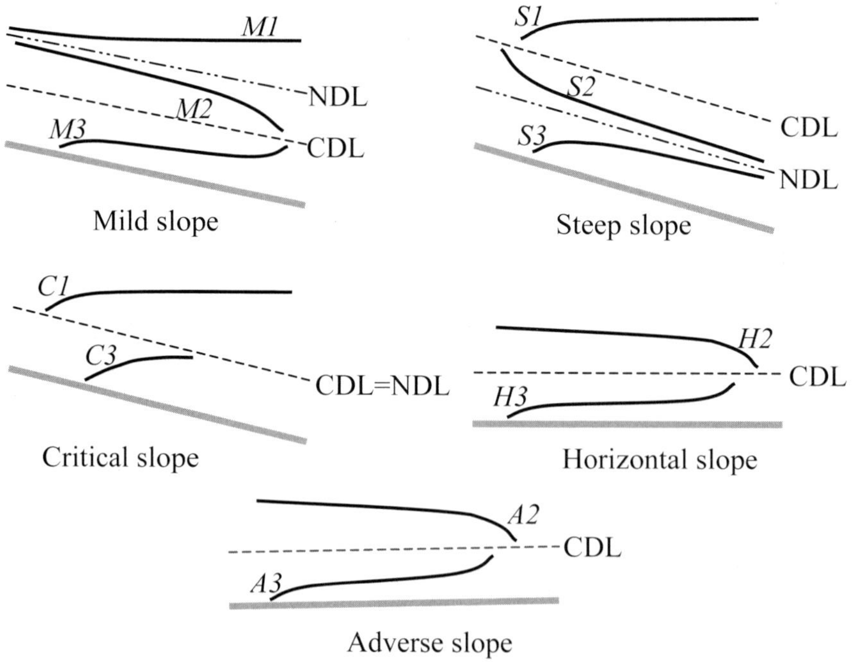

Figure 3-3: Gradually varied flow classifications (adapted from Chow, 1959)

3.6.3 Mixed Flow Profiles

In many systems, flow may transition between subcritical and supercritical regimes, the normal depth may change as a result of changes to grade and channel geometry, and critical depth may fluctuate based on changes in cross-sectional shape. The result is a variety of possible mixed, gradually varied flow profiles, as illustrated in Figure 3-4. In this regard, derivation of the entire profile assumes that each channel segment is sufficiently long to allow development of each individual water surface shape. In addition to a variety of gradually varied flow profiles, real flows may also transition to or from rapidly varied flow and may alternate between pressurized and free surface flow. In these cases, the different types and regimes of flow cannot be analyzed together; it is generally necessary to successively examine flow in separate segments.

3.6.4 Direct Step Method

There are two common techniques for quantitatively evaluating water surface profiles. The first, called the direct step method, solves the gradually varied flow equation for the distance over which a specified change in depth occurs (i.e., distance from depth). As a result, this method is ideally suited to prismatic channels since the channel cross section is independent of location within the channel. Equation 3-22 can be rewritten in finite difference form as

$$x_{i+1} - x_i = \frac{E_{i+1} - E_i}{S_o - S_f} \tag{3-24}$$

where the subscripts i and $i+1$ are used to identify cross sections, with i increasing in the downstream direction and S_f can be evaluated through the Chezy, Manning, or Darcy-Weisbach equations. For example, using the Manning equation,

$$S_f = \frac{n^2 V^2}{K_m^2 R^{4/3}} \tag{3-25}$$

Gradually varied flow computations must start from a control section, where a unique relationship between discharge and depth occurs (e.g., occurrence of critical or normal depth and flow at hydraulic structures or controls). For supercritical flow computations that proceed in the downstream direction, no modification to Equation 3-24 is necessary. However, for subcritical flows that proceed in the upstream direction, the numerator of the right-hand side should be multiplied by -1 to indicate that the current value of specific energy is subtracted from its assumed value in the upstream direction. This is a consistent sign convention since it will yield a negative value for Δx (i.e., $x_{i+1} - x_i$), which denotes upstream distance. Once the location of the unknown section is determined, it serves as the starting point for the next depth increment. Then, by successively increasing or decreasing the flow depth and determining where these depths occur, the entire water surface profile can be computed. It is important to note that large depth intervals should not be specified since friction slope is assumed to be identical to that for uniform flow.

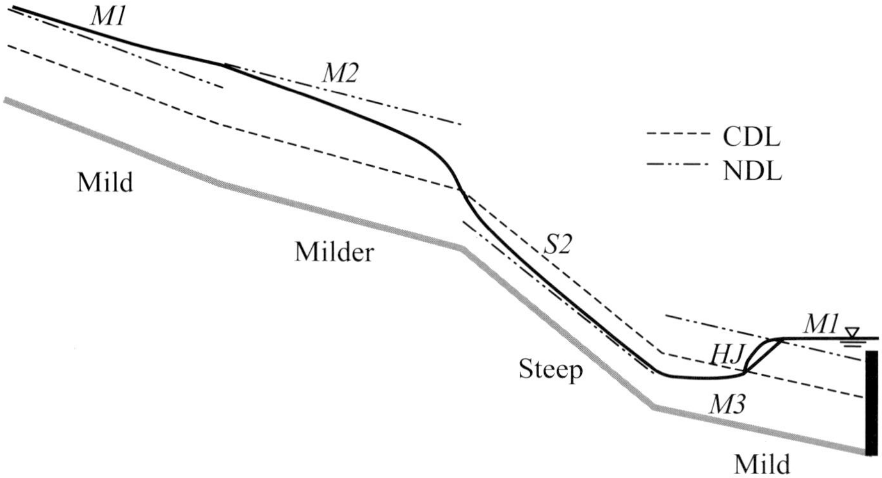

Figure 3-4: Sample mixed flow profile

3.6.5 Standard Step Method

While the direct step method is straightforward, it does not directly yield depth at the end of a known channel length. The standard step method overcomes this limitation by using energy principles to iteratively solve for the unknown depth at one end of the channel, assuming the depth at the other end is known (i.e., depth from distance). In addition, the method can easily be applied to any channel, not just those that are prismatic. Writing Equation 3-4 between two channel sections, accounting for channel slope and friction losses, yields

$$E_i + S_o \Delta x = E_{i+1} + S_f \Delta x \tag{3-26}$$

Similar to the direct step procedure, computations proceed downstream for supercritical flows and upstream for subcritical flows. For the subcritical case, consider that all information is known at section $i+1$. The method then begins by assuming a depth at i and computing the associated values for E_i and S_f. A check is then made to determine whether Equation 3-26 is satisfied (i.e., convergence). If it is not, a new value for y_i is assumed, and the process is repeated. Once the relationship is satisfied, the newly determined depth becomes the known value for the next channel segment. Because of the repetitive aspects of the solution, computations are often performed using computer spreadsheets or dedicated software packages.

3.6.6 Form Losses

As written, Equation 3-26 includes only energy losses caused by friction. However, in both pressurized and gravity sewers, additional form, or minor, losses are likely to occur at manholes, bends, contractions, expansions, and other appurtenances as a result of the viscous dissipation of turbulence created at these locations. When considered, the energy equation can be rewritten as

$$E_i + S_o \Delta x = E_{i+1} + S_f \Delta x + \sum h_l \tag{3-27}$$

where h_l represents individual form loss components, which can be estimated using the formulae listed in Table 3-3. For a more detailed analysis of form losses or for additional guidance on selection of form loss coefficients, the reader may consult references such as Chow (1959), Brown et al. (1996), Sangster (1958), or Crowe et al. (2001).

3.6.7 Hydraulic Gradient Evaluation

To analyze the hydraulic effectiveness of a preliminary design or existing sewer system, it is necessary to evaluate the hydraulic gradient, or HGL. The purpose is to ensure that the HGL elevation does not exceed the ground surface elevation at manholes and inlets. Such a condition would imply that flow exits the system at undesirable locations, causing potential flooding, and that design modifications are necessary.

An evaluation of the HGL is performed using energy principles, incorporating both friction and form losses, and typically begins at the system outfall. If the elevation of the downstream water surface, also known as the tailwater elevation, exceeds the downstream crown elevation of the sewer, the pipe is assumed to flow full and the HGL is set equal to the tailwater. If the tailwater is lower than the pipe's crown elevation, the pipe flows partially full and gravity flow prevails. For subcritical conditions, if the tailwater exceeds the conduit's critical depth, investigation by the Federal Highway Administration indicates that the HGL should be set equal to the greater of the tailwater or $(D + y_c)/2$ (Brown et al., 1996). Note that if supercritical conditions exist, flow is not influenced by downstream conditions; the analysis should, therefore, begin at an upstream control depth.

Table 3-3: Formulae for estimating form losses

Appurtenance	Formula	Comments
Expansion:		
Gravity flow	$h_l = 0.2\left(\frac{V_1^2 - V_2^2}{2g}\right)$	For expansions through straight-through manholes and similar transitions
Pressurized flow	$h_l = K_e\left[\frac{(V_1 - V_2)^2}{2g}\right]$	K_e ranges from 1.0 for an abrupt expansion to 0.2 for a well designed, gradual transition
Contraction:		
Gravity flow	$h_l = 0.1\left(\frac{V_2^2 - V_1^2}{2g}\right)$	For contractions through straight-through manholes and similar transitions
Pressurized flow	$h_l = K_c\left(\frac{V_2^2}{2g}\right)\left(1 - \frac{A_2}{A_1}\right)^2$	K_c ranges from 0.5 for an abrupt contraction to 0.1 for a well designed, gradual transition
Bend:	$h_l = 0.25\sqrt{\frac{\Phi}{90}}\left(\frac{V^2}{2g}\right)$	For central bend angle, Φ, of 40° or less; refer to Sangster (1958) for larger angles
Pipe junction (without manhole)	$h_l = \frac{Q_2V_2 - Q_1V_1 - Q_LV_L\cos\phi}{0.5g(A_2 + A_1)} + \frac{V_1}{2g} - \frac{V_2}{2g}$	Subscript L refers to lateral sewer flows, and ϕ is the interior angle between inflow and outflow mains
Manholes:		
Straight- or angled-through (two-way junction)	$h_l = K_m\frac{V^2}{2g}$	K_m ranges from 0.15 for straight-through to 0.75 for 45° angled-through to 1.0 for 90° angled-through manholes
Three –way junction	$h_l = K_j\frac{V^2}{2g}$	Refer to Sangster (1958) for values of K_j based on flow, junction geometry and relative diameters
Exit	$h_l = \frac{V^2}{2g}$	

Note: Subscripts 1 and 2 refer to upstream and downstream (i.e., inflow and outflow) cross sections of sewer, respectively
Source: AISI (1990) and Brown et al. (1996)

Once the elevation of the HGL at the downstream end is known, the analysis continues methodically through the entire system, proceeding on the basis of one length of conduit at a time. When the HGL exceeds the pipe crown elevation, a surcharge condition exists and pressure flow computations are indicated; if it falls below the crown, open channel calculations (e.g., standard step method) should be used.

3.7 *RAPIDLY VARIED FLOWS*

When rapid changes in flow depth occur over short distances, pressure distributions cannot be assumed to be hydrostatic and velocity distributions become highly nonuniform. The flow profile may be continuous or discontinuous. Such flow conditions are referred to as rapidly varied. Common examples are flow through channel transitions, flow over weirs or spillways, and hydraulic jumps. In these cases, large energy losses are incurred that cannot be reliably analyzed by using energy principles alone.

3.7.1 Hydraulic Jump in Sewer Systems

When supercritical flow enters a reach in which flow is subcritical (e.g., shift from steep to mild slope), an abrupt transition in depth occurs. Known as a hydraulic jump (HJ), flow at this point will pass through critical depth and is characterized by the formation of a rolling, undular wave (see Figure 3-4). Applying momentum principles to a control volume containing the jump yields

$$\frac{Q^2}{gA_1} + A_1\overline{y_1} = \frac{Q^2}{gA_2} + A_2\overline{y_2} \tag{3-28}$$

where the subscripts *1* and *2* denote conditions upstream and downstream of the jump, respectively, and $\overline{y}$ is distance below the free surface to the centroid of the corresponding flow sections. The values of y_1 and y_2 that implicitly satisfy the equation are referred to as sequent depths, and the total term on either side of Equation 3-28 is called specific momentum, *M*. Formulae for *M* for various channel shapes are presented in Table 3-4. Thus, knowing one sequent depth, the relationships can be used to implicitly evaluate the other.

For rectangular channels, Equation 3-28 can be manipulated to provide an explicit relationship between sequent (conjugate) depths, expressed as

$$\frac{y_1}{y_2} = \frac{1}{2}\left(-1 + \sqrt{1 + 8Fr_2^2}\right) \tag{3-29}$$

or

$$\frac{y_2}{y_1} = \frac{1}{2}\left(-1 + \sqrt{1 + 8Fr_1^2}\right) \qquad (3\text{-}30)$$

Table 3-4: Specific momentum for various cross sections

Shape	Specific momentum (M)
Rectangular 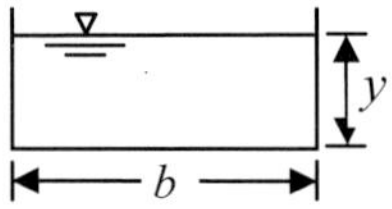	$\frac{by^2}{2} + \frac{Q^2}{gby}$
Trapezoidal[a] 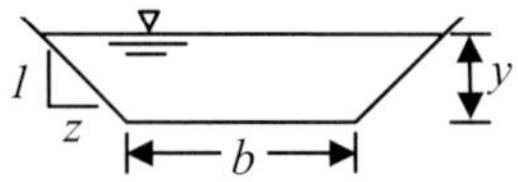	$\frac{by^2}{2} + \frac{zy^3}{3} + \frac{Q^2}{gy(b+zy)}$
Parabolic[b] 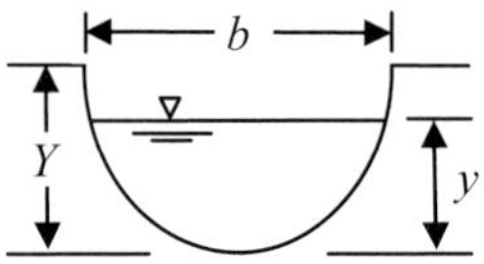	$\frac{4}{15}\frac{by^{5/2}}{\sqrt{Y}} + \frac{1.5Q^2\sqrt{Y}}{gby^{3/2}}$
Circular[b] 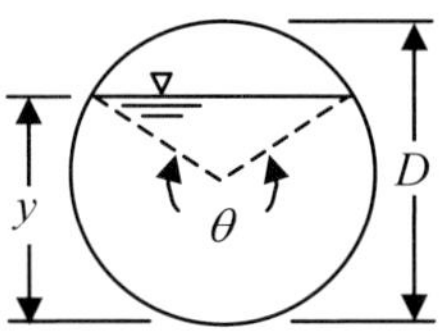	$\frac{D^3}{24}\left[3\sin\left(\frac{\theta}{2}\right) - \sin^3\left(\frac{\theta}{2}\right) - \frac{3\theta}{2}\cos\left(\frac{\theta}{2}\right)\right] + \frac{Q^2}{gD^2(\theta - \sin\theta)/8}$

[a] For triangular sections, use $b = 0$

[b] use θ in radians, where $\theta = 2\cos^{-1}\left(1 - \frac{2y}{D}\right)$

Source: Adapted from Sturm (2001)

The length of a hydraulic jump, L_j, is somewhat difficult to define. It is not clear where to mark its beginning and end because of the highly turbulent flow surface and air entrainment. Nevertheless, experimental results by the U.S.

Bureau of Reclamation for rectangular channels indicate that L_j varies with the upstream Froude number, Fr_1, and downstream subcritical depth, y_2, as indicated in Table 3-5. In the absence of adequate experimental data, values provided can be used to approximate jump lengths in other cross-sectional shapes.

Table 3-5: Approximate lengths of hydraulic jumps

Fr_1	$\frac{L_j}{y_2}$
2	4.4
3	5.25
4	5.8
5	6.0
6	6.1
7	6.15
8	6.15

Source: Adapted from USBR (1955)

As previously noted, standard step computations proceed upstream for subcritical (i.e., downstream controlled) flow and downstream for supercritical (i.e., upstream controlled) flow. However, a hydraulic jump is capable of serving as a control for both upstream and downstream flows. This fact will often become evident as a result of nonconvergence of the standard step method for the reach in which the jump is located. The location of the jump, therefore, must be determined by matching the gradually varied flow profile depths immediately before and after the jump with corresponding sequent depths.

Hydraulic jumps undergo significant mixing and turbulence, which are associated with high energy loss, and can be highly erosive. As a result, they should either be prevented from occurring in sewer systems or their location should be predicted in order to provide additional protection for the conduit.

3.8 SOLVED PROBLEMS

Problem 3.1 Continuity

Flow travels through a 12-in diameter sewer at 8 fps. If the pipe expands to a diameter of 18 in, what are the velocity and volumetric flow rate in the larger section?

Solution

From Equation 3-2,

$$V_{18} = V_{12}\frac{A_{12}}{A_{18}} = V_{12}\left(\frac{D_{12}}{D_{18}}\right)^2 = 8\left(\frac{12}{18}\right)^2 = \underline{\underline{3.56\ fps}}$$

and, by definition,

$$Q_{18} = Q_{12} = V_{12}A_{12} = 8\left(\frac{\pi(1)^2}{4}\right) = \underline{\underline{6.28\ cfs}}$$

Problem 3.2 Continuity and flow classification

A trapezoidal channel having a bottom width of 1.5 m and side slopes of 2H:1V carries a discharge of 0.5 m^3/s. If the depths at two adjacent cross sections are measured as 0.25 m and 0.35 m, respectively, determine the velocities at each section and classify the flow in the channel reach as subcritical or supercritical.

Solution

Using the area of a trapezoidal section, given in Table 3-1, and noting that $Q_1 = Q_2$,

$$V_1 = \frac{Q}{A_1} = \frac{0.5}{[1.5 + 2(0.25)]0.25} = \underline{\underline{1.0\ \frac{m^3}{s}}}$$

and

$$V_2 = \frac{Q}{A_2} = \frac{0.5}{[1.5 + 2(0.35)]0.35} = \underline{\underline{0.65\ \frac{m^3}{s}}}$$

Hydraulic depth, D, is defined as the ratio of area to top width. For this trapezoidal channel,

$$D_1 = \frac{[1.5 + 2(0.25)]0.25}{1.5 + 2(2)(0.25)} = 0.20\ m$$

Then using Equation 3-7,

$$Fr_1 = \frac{V_1}{\sqrt{gD_1}} = \frac{1.0}{\sqrt{9.81 \times 0.20}} = 0.71\ m$$

Using a similar analysis, the Froude number at section *2* is 0.40. Since $Fr < 1.0$ at both sections, flow in the reach is characterized as <u>subcritical</u>.

Problem 3.3 Energy

Flow travels under pressure at a rate of 1.0 cfs through a 6-in diameter, cast iron pipe. Reach *AB* of the pipe has a length of 1,000 ft, with section *A* being 100 ft higher in elevation than section *B*. If energy losses are approximated as 20 ft, what is the pressure difference between sections *A* and *B*? Assume $\nu = 9.3 \times 10^{-6}$.

Solution

The velocity is computed as

$$V = \frac{Q}{A} = \frac{1}{\left(\frac{\pi(0.5)^2}{4}\right)} = 5.09\ fps$$

and the energy equation written between *A* and *B* is expressed as

$$\frac{p_A}{\gamma} + z_A + \frac{V_A^2}{2g} = \frac{p_B}{\gamma} + z_B + \frac{V_B^2}{2g} + h_L$$

Realizing that $V_1 = V_2$,

$$p_B - p_A = \gamma(z_A - z_B - h_L) = 62.4(100 - 20) = 4{,}992\ psf = \underline{\underline{34.7\ psi}}$$

Problem 3.4 HGL/EGL

Sketch HGL and EGL for pipe system shown in Figure P3-4a.

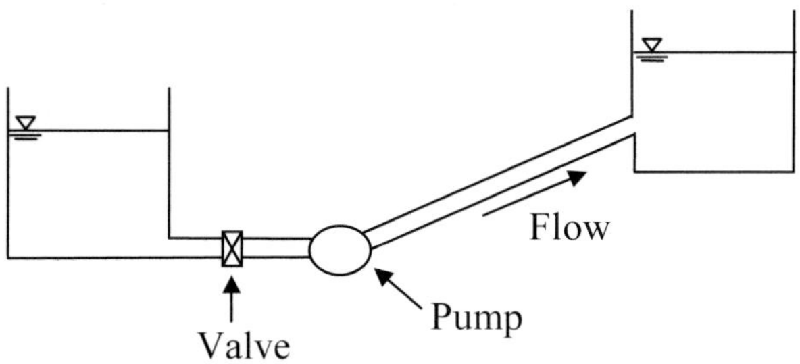

Figure P3-4a

Solution

The HGL and EGL are drawn in Figure P3-4b. Notable features include the (a) form losses that occur at the pipe entrance and exit and at the valve; (b) energy added through pressure by the pump; and (c) constant slope of the HGL and EGL due to a constant pipe diameter, and thus velocity.

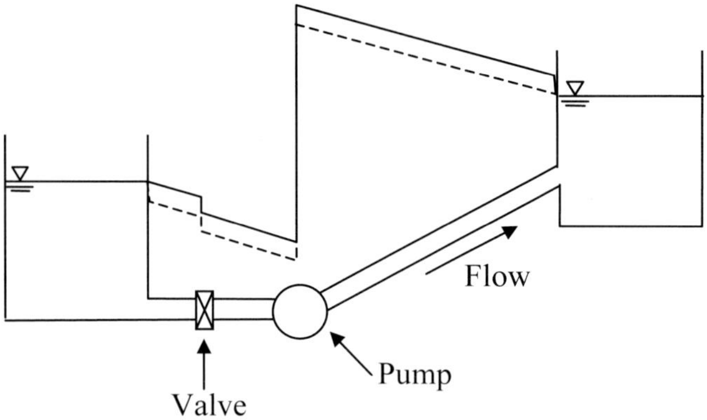

Figure P3-4b

Problem 3.5 HGL/EGL

Shown in Figure P3-5 are the HGL and EGL for a particular flow system. (a) Identify the flow direction and distinguish between the HGL and EGL. (b) Is there a reservoir or storage tank located in the system? If so, where? (c) Do all the pipes in this system have the same diameter? If not, which is the smallest? (d) What exists at *B*? (e) What likely exists at *D*?

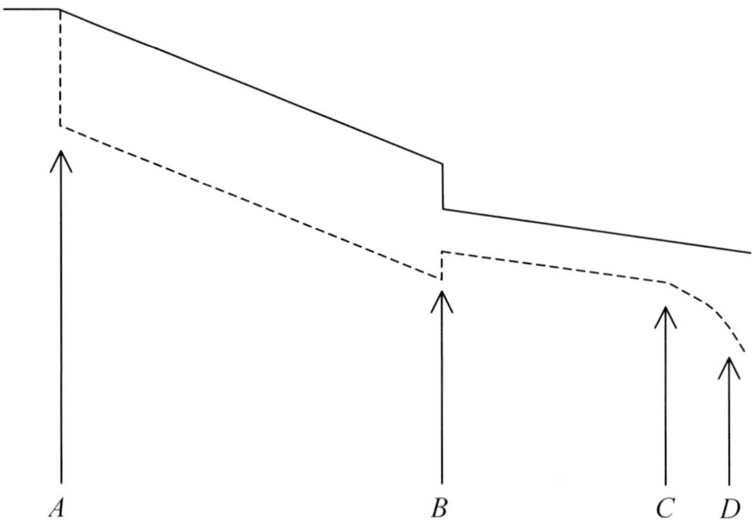

Figure P3-5

Solution

(a) The solid line is the EGL, indicating total energy, and the HGL is the dashed line. As a result of head losses, the lines must have a negative slope; thus flow is from left to right.
(b) A reservoir or storage tank exists prior to *A*, where the HGL and EGL coincide, thus indicating a zero velocity.
(c) The distance between the HGL and EGL is greatest between sections *A* and *B*, thus indicating the largest velocity. A larger velocity implies a smaller diameter; therefore section *A* and *B* has the smallest diameter. In addition, the slope of lines between *A* and *B* is steeper than in other portions of the system, indicating a larger head loss. Considering available resistance equations, it is obvious that larger losses are associated with higher velocities.
(d) An abrupt expansion exists at *B*. It can be characterized through the associated drop in the EGL (i.e., form loss) and abrupt drop in velocity head.
(e) Since the lines gradually diverge, indicating a slowly increasing velocity, a gradual contraction or nozzle exists at *D*.

Problem 3.6 Momentum

A pressurized sewer contains a 45° horizontal expansion bend (see Figure P3-6) which transitions from a 12-in diameter to 18-in diameter. The total loss through the bend is 3 ft, and the sewer carries a discharge of 20 cfs. If the pressure at the inlet to the bend is 23 psi, find the resultant force required to stabilize the bend.

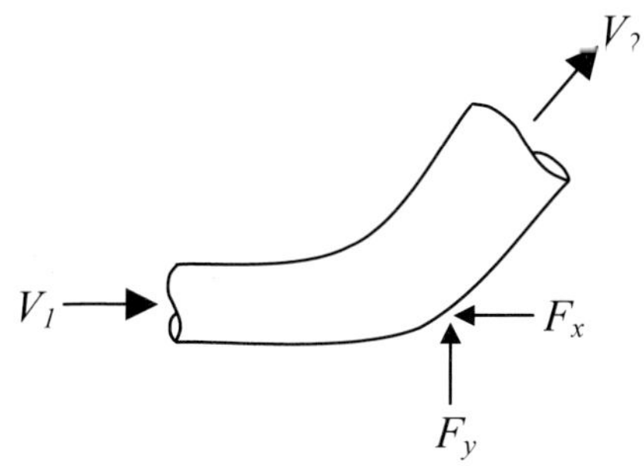

Figure P3-6

Solution

The upstream and downstream velocities are computed as

$$V_1 = \frac{Q}{A_1} = \frac{20}{\left(\frac{\pi(1.0)^2}{4}\right)} = 25.46 \ fps$$

and

$$V_2 = V_1 \frac{A_1}{A_2} = V_1 \left(\frac{D_1}{D_2}\right)^2 = 25.46\left(\frac{12}{24}\right)^2 = 6.37 \ fps$$

Noting that this is a horizontal bend, the energy equation can be expressed as

$$\frac{p_1}{\gamma} + \frac{V_1^2}{2g} = \frac{p_2}{\gamma} + \frac{V_2^2}{2g} + h_L$$

which can be solved for pressure at the bend exit,

$$p_2 = \gamma\left(\frac{p_1}{\gamma} + \frac{V_1^2 - V_2^2}{2g} - h_L\right) = 62.4\left(\frac{23 \times 144}{62.4} + \frac{(25.46)^2 - (6.37)^2}{2 \times 32.2} - 3\right)$$

or p_2 = 3,714 psf. The momentum equation is applied in both the x and y directions to determine component forces. For the x direction,

$$p_1 A_1 - \cos 45 \, p_2 A_2 - F_x = \rho Q[\cos 45 V_2 - V_1]$$

or

$$F_x = (23 \times 144)\left(\frac{\pi(1.0)^2}{4}\right) - (\cos 45)(3714)\left(\frac{\pi(2.0)^2}{4}\right)$$
$$- (1.94 \times 20)[(\cos 45)(6.37) - 25.46] = -4{,}836 \ lbs$$

The negative sign indicates that the wrong direction was initially assumed; the force instead acts to the right. For the y direction,

$$F_y - \sin 45 \, p_2 A_2 = \rho Q[\sin 45 V_2 - 0]$$

or

$$F_y = (\sin 45)(3714)\left(\frac{\pi(2.0)^2}{4}\right)$$
$$+ (1.94 \times 20)[(\cos 45)(6.37) - 0] = 8{,}425 \ lbs$$

The resultant force is computed as

$$F_R = \sqrt{(4836)^2 + (8425)^2} = \underline{\underline{9{,}714\ lbs}}$$

acting upward and to the right at an angle of $tan^{-1}\left(\dfrac{8425}{4836}\right) = 60^\circ \angle$

Problem 3.7 Momentum

Flow travels under a sluice gate installed in a rectangular channel as shown in Figure P3-7. At a moderate distance upstream and downstream of the gate, depths are 3 m and 0.5 m, respectively, and pressure distributions can be assumed to be hydrostatic. If the discharge is 3.5 m^3/s and energy losses are considered negligible, compute the force per unit width, F_g, required to hold the gate in place.

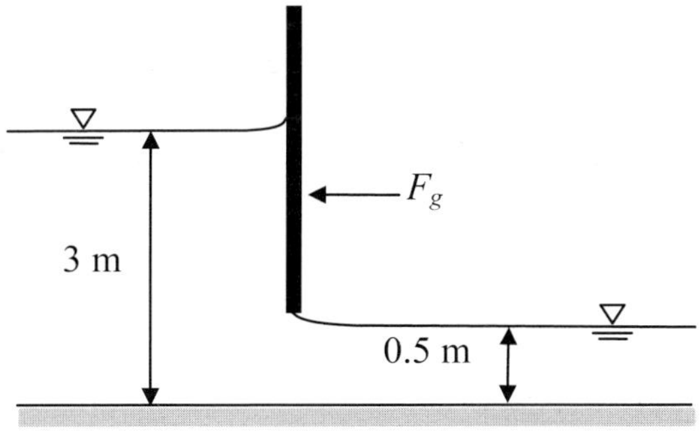

Figure P3-7

Solution

The upstream and downstream velocities are computed as

$$V_1 = \frac{Q}{A_1} = \frac{3.5}{(1 \times 3)} = 1.17\frac{m}{s}$$

and

$$V_2 = \frac{Q}{A_2} = \frac{3.5}{(1 \times 0.5)} = 7.0\frac{m}{s}$$

The momentum equation, as follows, is applied in the x direction to determine F_g:

$$\gamma \overline{y_1} A_1 - \gamma \overline{y_2} A_2 - F_g = \rho Q (V_2 - V_1)$$

or

$$F_g = (9810)\left(\frac{3}{2}\right)(3 \times 1) - (9810)\left(\frac{0.5}{2}\right)(0.5 \times 1) - (1000)(3.5)(7.0 - 1.17) = \underline{\underline{22{,}514 \ N}}$$

Problem 3.8 Critical depth

Determine the critical depth in a 1.6-m wide rectangular channel carrying 1.8 m^3/s.

Solution

Applying Equation 3-7 with $y = y_c$ (i.e., $Fr = 1.0$) and noting that hydraulic depth, D, for a rectangular cross section is equivalent to depth yields

$$V = \sqrt{g y_c}$$

or in terms of a unit width flow rate, q (i.e., (Q/B)),

$$\frac{q}{y_c} = \sqrt{g y_c}$$

where q is equivalent to 1.8/1.6, or 1.125 m^3/s/m. Solving for y_c provides a convenient equation for finding critical depth in a rectangular channel.

$$y_c = \sqrt[3]{\frac{q^2}{g}} = \sqrt[3]{\frac{(1.13)^2}{9.81}} = \underline{\underline{0.51\ m}}$$

Problem 3.9 Critical depth

Determine the critical depth in a trapezoidal channel having a bottom width of 6 ft and 1H:1V side slopes. The channel carries a discharge of 150 cfs.

Solution

Applying Equation 3-6 and using geometric properties listed in Table 3-1 gives

$$\frac{BQ^2}{gA^3} = 1 = \frac{(6 + 2y_c)(150)^2}{(32.2)(6y_c + y_c^2)^3}$$

Solving by trial-and-error, numerical analysis, or computational solver yields y_c = $\underline{\underline{2.34\text{ ft}}}$.

Problem 3.10 Critical depth

Flow travels through a 36-in diameter concrete sewer at a rate of 8.8 cfs and a depth of 15 in. Is flow supercritical or subcritical? What is the critical depth?

Solution

Equation 3-6 can be used to find the Froude number as follows:

$$\theta = 2\cos^{-1}\left(1 - \frac{2y}{D}\right) = 2\cos^{-1}\left(1 - \frac{2(1.25)}{3}\right) = 2.81\,rad$$

$$A = \frac{(\theta - \sin\theta)D^2}{8} = \frac{(2.81 - \sin 2.81)(3)^2}{8} = 2.80\ ft^2$$

$$B = (3)\sin\frac{2.81}{2} = 2.96\ ft$$

$$Fr = \frac{BQ^2}{gA^3} = \frac{(2.96)(8.8)^2}{(32.2)(2.8)^3} = 0.32$$

Since $Fr < 1.0$, flow is <u>subcritical</u>. The critical depth can be determined solving Equation 3-6, for when Fr is 1.0, or can by approximated by Equation 3-8.

$$y_c = \left(\frac{1.01}{D^{0.26}}\right)\left(\frac{Q^2}{g}\right)^{0.25} = \left(\frac{1.01}{(3)^{0.26}}\right)\left(\frac{(8.8)^2}{32.2}\right)^{0.25} = 0.945\ ft = \underline{\underline{11.3\ in}}$$

Notice that $y_c < y$, thus confirming that flow is subcritical.

Problem 3.11 Specific Energy

The channel described in Problem 3.8 carries a discharge of 1.8 m^3/s at a depth of 0.85 m. The channel undergoes a smooth transition in which the channel bottom rises 15 cm. Assuming friction losses are negligible, determine the depth of flow at the transition and the associated change in water surface elevation.

Solution

Conservation of energy through the transition can be expressed using specific energy (i.e., Equation 3-4). Specific energy just before the transition can be evaluated as

$$E_1 = y_1 + \frac{Q^2}{2gA_1^2} = 0.85 + \frac{(1.8)^2}{2(9.81)(1.6 \times 0.85)^2} = 0.94\ m$$

Then, by definition,

$$E_1 = E_2 + \Delta z$$

or

$$E_2 = E_1 - \Delta z = 0.94 - 0.15 = 0.79\ m$$

Specific energy at the transition can be expressed as

$$E_2 = y_2 + \frac{Q^2}{2gA_2^2} = 0.79 = y_2 + \frac{(1.8)^2}{2(9.81)(1.6 \times y_2)^2}$$

or, alternatively,

$$0 = y_2^3 - 0.79 y_2^2 + 0.0645$$

Solution of the cubic equation yields three possible depths (i.e., roots) that satisfy the expression. One is negative, which is not physically possible. The remaining two are 0.41 m and 0.62 m. Notice that when compared to the critical depth of 0.51 derived in Problem 3.8, these roots represent alternate depths (see Figure 3-2); thus one is supercritical and the other is subcritical. Since upstream flow is subcritical (i.e., 0.85 m > 0.51 m), the supercritical solution is not likely because flow would have to pass through critical depth at the transition. Therefore, the depth at the transition is 0.62 m. The associated water level downstream is 0.62 m + 0.15 m, or 0.77 m. Thus, the change in water surface elevation through the transition is 0.85 m – 0.77 m, or a 0.08 m drop in elevation.

Problem 3.12 Specific Energy

For Problem 3.11, determine the maximum height of transition that could occur before upstream conditions are affected (i.e., choked).

Solution

For this case, the value of E_2 cannot drop below that of E_{min} without affecting upstream conditions. This is evident by examining Figure 3-2; if $E_2 < E_{min}$, a new specific energy curve is created, indicating that flow is decreased. To solve for the maximum transition height, let $y_2 = y_c = 0.51$ m, and compute the associated specific energy as

$$E_{min} = E_2 = y_2 + \frac{Q^2}{2gA_2^2} = 0.51 + \frac{(1.8)^2}{2(9.81)(1.6 \times 0.51)^2} = 0.76\ m$$

Then the maximum transition is

$$\Delta z_{max} = E_1 - E_2 = 0.94 - 0.76 = \underline{\underline{0.18\ m}}$$

Problem 3.13 Specific Energy

If the channel in Problem 3.11 contracts to a width of 1.3 m, without a transition in invert elevation, what is the depth of flow in the contraction and associated change in water surface elevation? Assume friction losses are negligible.

Solution

Using results from Problem 3.11, conservation of energy implies that

$$E_1 = E_2 = 0.94$$

Then

$$E_2 = y_2 + \frac{Q^2}{2gA_2^2} = 0.94 = y_2 + \frac{(1.8)^2}{2(9.81)(1.3 \times y_2)^2}$$

or alternatively,

$$0 = y_2^3 - 0.94 y_2^2 + 0.098$$

which has three roots (i.e., -0.28 m, 0.44 m, and 0.78 m). Since upstream flow is subcritical, the correct root is $\underline{\underline{0.78\ m}}$. The associated water level change is 0.85 m – 0.78 m, or a $\underline{\underline{0.07\ m\ drop}}$ in elevation.

Problem 3.14 Specific Energy

If the channel in Problem 3.11 is subjected to both a 15 cm change in bed elevation and a contraction to a width of 1.3 m, determine the depth of flow at the transition. Assume friction losses are negligible.

Solution

Applying energy principles,

$$E_2 = E_1 - \Delta z = 0.94 - 0.15 = 0.79\ m$$

and

$$E_2 = y_2 + \frac{Q^2}{2gA_2^2} = 0.79 = y_2 + \frac{(1.8)^2}{2(9.81)(1.3 \times y_2)^2}$$

which can be rewritten as

$$0 = y_2^3 - 0.79 y_2^2 + 0.098$$

However, this equation has no positive root. To investigate further, consider the minimum energy at the transition by letting $y_2 = y_c$, where

$$y_c = \sqrt[3]{\frac{q^2}{g}} = \sqrt[3]{\frac{\left(1.8/1.3\right)^2}{9.81}} = 0.58\ m$$

The minimum energy level is computed as

$$E_{min} = y_c + \frac{Q^2}{2gA_c^2} = 0.58 + \frac{(1.8)^2}{2(9.81)(1.3 \times 0.58)^2} = 0.87\ m$$

Notice that E_2 computed previously (i.e., 0.79 m) is less than the minimum level of 0.87 m, a condition that is not possible without changing the flow rate. Therefore, critical depth will occur at the prescribed transition and upstream flow will be choked. An alternative way to derive this conclusion is by considering the maximum transition height. Consider that

$$\Delta z_{max} = 0.94 - 0.87 = 0.07\ m$$

which implies that the 0.15 m transition specified in the problem statement is greater than the maximum.

Comments: The inclusion of either the invert transition or the contraction alone does not affect upstream conditions. However, when both are considered, upstream flow is choked.

3.15 Specific Energy

A 9-ft diameter circular pipe carries 260 cfs at a depth of 7 ft. A subcritical transition is installed at the end of the pipe and consists of a 3 ft drop in the channel invert leading to a 6.5-ft wide rectangular channel. Assuming energy losses are negligible, compute the depth in the downstream rectangular channel and the change in water surface elevation within the transition.

Solution

For the upstream pipe,

$$\theta = 2\cos^{-1}\left(1 - \frac{2y}{D}\right) = 2\cos^{-1}\left(1 - \frac{2(7)}{9}\right) = 4.32\ rad$$

and

$$A = \frac{(\theta - \sin\theta)D^2}{8} = \frac{(4.32 - \sin 4.32)(9)^2}{8} = 53.10\ ft^2$$

The specific energy just before the transition is evaluated as

$$E_1 = y_1 + \frac{Q^2}{2gA_1^2} = 7 + \frac{(260)^2}{2(32.2)(53.10)^2} = 7.37\ ft$$

Then,

$$E_1 + \Delta z = E_2 = 7.37 + 3 = 10.37\ ft$$

Specific energy at the transition is expressed by

$$E_2 = y_2 + \frac{Q^2}{2gA_2^2} = 10.37 = y_2 + \frac{(260)^2}{2(32.2)(6.5 \times y_2)^2}$$

or alternatively,

$$0 = y_2^3 - 10.37 y_2^2 + 24.84$$

Solving for the subcritical root yields a depth of 10.08 ft. The water surface elevation upstream is 7 ft + 3 ft, or 10 ft. Thus, the change in water surface elevation through the transition is 10.08 ft – 10 ft, or a 0.08 ft rise in elevation.

Problem 3.16 Flow resistance (Chezy)

Use the Chezy equation to determine the velocity and flow rate through a 500-ft length of 24-in diameter cast iron pipe flowing under pressure if friction losses total 5 ft. Assume $\nu = 1.2 \times 10^{-5}$ ft²/s.

Solution

Friction slope, S_f, is equivalent to 5/500, or 0.01 ft/ft, and for a full flowing pipe, $R = D/4$, or 0.5 ft. Assuming $n = 0.013$ and applying Equation 3-10 for the Chezy coefficient yields

$$C = \frac{41.65 + \frac{0.00281}{0.01} + \frac{1.811}{0.013}}{1 + 0.013\left(41.65 + \frac{0.00281}{0.01}\right)(0.5)^{-1/2}} = 102.34$$

The Chezy equation can then be used to evaluate velocity.

$$V = C\sqrt{RS_f} = 102.34\sqrt{0.5 \times 0.01} = \underline{\underline{7.24\ fps}}$$

The corresponding discharge is

$$Q = VA = 7.24\left(\frac{\pi(2)^2}{4}\right) = \underline{\underline{22.75\ cfs}}$$

Problem 3.17 Flow resistance (Darcy-Weisbach)

Solve Problem 3.16 using the Darcy-Weisbach equation.

Solution

Assume $k_s = 8.5 \times 10^{-4}$ ft for cast iron. The solution is iterative since both f and V are unknown. Equation 3-14 is expressed as

$$V = \sqrt{\frac{8gRS_f}{f}} = \sqrt{\frac{(8)(32.2)(0.5)(0.01)}{f}} = \frac{1.135}{\sqrt{f}}$$

and the Reynolds number can be evaluated by

$$Re = \frac{V4R}{\upsilon} = \frac{VD}{\upsilon} = \frac{V(2)}{1.2 \times 10^{-5}} = 1.67 \times 10^5 V$$

Begin by assuming $f = 0.025$; then $V = 7.18$ fps and $Re = 1.2 \times 10^6$. Using Equation 3-18, f is checked, as follows:

$$f = \frac{1.325}{\left[ln\left(\frac{k_s}{3.7D} + \frac{5.74}{Re^{0.9}}\right)\right]^2} = \frac{1.325}{\left[ln\left(\frac{8.5 \times 10^{-4}}{3.7D} + \frac{5.74}{\left(1.2 \times 10^6\right)^{0.9}}\right)\right]^2} = 0.017$$

Since the computed friction factor does not match the assumed value, a second iteration is necessary. Assuming $f = 0.017$, $V = 8.71$ fps and $Re = 1.45 \times 10^6$. Recomputing f yields 0.017, which matches the assumed value. Therefore, the velocity is $\underline{\underline{8.71\text{ fps}}}$, and

$$Q = VA = 8.71\left(\frac{\pi(2)^2}{4}\right) = \underline{\underline{27.36\ cfs}}$$

Comments: Using the Darcy-Weisbach equation results in a discharge that is approximately 20 percent greater than that derived from the Chezy equation.

Problem 3.18 Flow resistance (Manning)

Solve Problem 3.16 using the Manning equation.

Solution

Assuming $n = 0.013$, Equation 3-12 yields

$$V = \frac{K_m}{n} R^{2/3} S_f^{\ 1/2} = \frac{1.49}{0.013}(0.5)^{2/3}(0.01)^{1/2} = \underline{\underline{7.20\ fps}}$$

The corresponding flow rate is

$$Q = VA = 7.20\left(\frac{\pi(2)^2}{4}\right) = \underline{\underline{22.62\ \ cfs}}$$

Comments: The result derived from the Manning equation correlates well with that derived from the Chezy equation, but is approximately 17 percent less than that computed with the Darcy-Weisbach equation.

Problem 3.19 Flow resistance (Hazen-Williams)

Solve Problem 3.16 using the Hazen-Williams equation.

Solution

Assuming $C_H = 130$, Equation 3-13 is used to evaluate velocity.

$$V = K_H C_H R^{0.63} S_f^{0.54} = (1.318)(130)(0.5)^{0.63}(0.01)^{0.54} = \underline{\underline{9.21\ fps}}$$

The associated flow rate is

$$Q = VA = 9.21\left(\frac{\pi(2)^2}{4}\right) = \underline{\underline{28.93\ \ cfs}}$$

Comments: Note that the result is within 5 percent of that derived using the Darcy-Weisbach equation, but is significantly greater than results from either the Chezy or Manning equations.

Problem 3.20 Flow resistance (Darcy-Weisbach)

Use the Darcy-Weisbach equation to estimate friction losses for the pipe described in Problem 3.3.

Solution

Assume $k_s = 8.5 \times 10^{-4}$ ft for cast iron pipe. For full pipe flow, $R = D/4$, so that the Reynolds number becomes

$$Re = \frac{VD}{\upsilon} = \frac{(5.09)(0.5)}{9.3 \times 10^{-6}} = 2.74 \times 10^{5}$$

From the Swamee-Jain equation,

$$f = \frac{1.325}{\left[ln\left(\frac{k_s}{3.7D} + \frac{5.74}{Re^{0.9}}\right)\right]^2} = \frac{1.325}{\left[ln\left(\frac{8.5 \times 10^{-4}}{3.7(0.5)} + \frac{5.74}{\left(2.74 \times 10^{5}\right)^{0.9}}\right)\right]^2} = 0.023$$

Solving Equation 3-14 for friction slope, noting that $R = D/4$ for full pipe flow, yields

$$S_f = \frac{fV^2}{D2g} = \frac{(0.023)(5.09)^2}{(0.5)(2)(32.2)} = 0.019\ ft/ft = \frac{h_f}{L}$$

Thus, for a 1,000 ft length of pipe, h_f is approximately 19 ft.

Problem 3.21 Flow resistance (Hazen-Williams)

Use the Hazen-Williams equation to estimate friction losses for the pipe described in Problem 3.3.

Solution

Assume $C_H = 130$. Solving Equation 3-13 for friction slope, noting that $R = D/4$ for full pipe flow, gives

$$S_f = \left(\frac{V}{K_H C_H R^{0.63}}\right)^{1.85} = \left(\frac{5.09}{(1.318)(130)(0.5/4)^{0.63}}\right)^{1.85} = 0.017\ ft/ft = \frac{h_f}{L}$$

For a 1,000 ft length of pipe, h_f is approximately 17 ft.

Problem 3.22 Flow resistance (Chezy)

Use the Chezy equation to estimate friction losses for the pipe described in Problem 3.3.

Solution

Assume $n = 0.013$. Substituting Kutter's equation into Equation 3-9 yields

$$V = \left(\frac{K_1 + \dfrac{K_2}{S_f} + \dfrac{K_3}{n}}{1 + n\left(K_1 + \dfrac{K_2}{S_f}\right)R^{-1/2}} \right)\sqrt{RS_f}$$

or

$$5.09 = \left(\frac{41.65 + \dfrac{0.00281}{S_f} + \dfrac{1.811}{0.013}}{1 + 0.013\left(41.65 + \dfrac{0.00281}{S_f}\right)\left(\dfrac{0.5}{4}\right)^{-1/2}} \right)\sqrt{\left(\frac{0.5}{4}\right)S_f}$$

Solving for S_f yields 0.041 ft/ft; thus h_f is approximately 41 ft for the 1,000 ft length of pipe.

Problem 3.23 Flow resistance (Manning)

Use the Manning equation to estimate friction losses for the pipe described in Problem 3.3.

Solution

Assuming $n = 0.013$, and rearranging Equation 3-12 yields

$$S_f = \left(\frac{Vn}{K_m R^{2/3}}\right)^2 = \left(\frac{(5.09)(0.013)}{(1.49)(0.5/4)^{2/3}}\right)^2 = 0.032\ ft/ft$$

So that h_f is approximately 32 ft for the 1,000 ft length of pipe

Comments: The results for the various resistance equations range from 17 ft to 41 ft. However, those associated with the Darcy-Weisbach and Hazen-Williams equation are more consistent with one another (i.e., 17 and 19 ft). As previously noted, these relationships are primarily intended for pressurized conduits, and therefore are likely to represent more realistic values for the current problem.

Problem 3.24 Flow resistance (Hazen-Williams)

Determine the diameter of galvanized iron pipe required to convey 0.2 m^3/s over a distance of 2,000 m. The pipe drops 50 m in elevation over the entire length, and the upstream and downstream pressures are 500 kPa and 420 kPa, respectively. Assume form losses are negligible.

Solution

Applying the energy equation over the 2,000 m length of pipe, noting that $V_1 = V_2$, yields

$$\frac{p_1}{\gamma} + z_1 = \frac{p_2}{\gamma} + z_2 + h_f$$

or

$$h_f = \frac{p_1 - p_2}{\gamma} + (z_1 - z_2) = \frac{500 - 420}{9.81} + 50 = 58.2\ m$$

Then S_f is equal to 58.2/2000 (i.e., h_f/L), or 0.029 m/m. Expressing Equation 3-13 in terms of flow rate and solving for diameter yields the following expression:

$$D = \left(\frac{Q}{0.104 K_H C_H \pi S_f^{0.54}} \right)^{0.38}$$

or, with $C_H = 135$,

$$D = \left(\frac{0.2}{(0.104)(0.849)(135)\pi(0.029)^{0.54}} \right)^{0.38} = 0.283\ m$$

The next largest commercially available pipe size, or 300 mm, should be used.

Comments: Since a larger pipe than required is being specified, the resulting effect (i.e., reduced friction loss and velocity) should be subsequently verified.

Problem 3.25 Uniform flow (Manning)

Use the Manning equation to compute the discharge in a 24-in diameter, cast iron pipe flowing at a depth of 5.5 in and laid on a slope of 0.004 ft/ft.

Solution

Assume $n = 0.013$. Using Table 3-1,

$$\theta = 2\cos^{-1}\left(1 - \frac{2y}{D}\right) = 2\cos^{-1}\left(1 - \frac{2(5.5)}{24}\right) = 2.0\ rad$$

$$A = \frac{(\theta - \sin\theta)D^2}{8} = \frac{(2.0 - \sin 2.0)(2)^2}{8} = 0.545\ ft^2$$

$$R = \frac{D}{4}\left(1 - \frac{\sin\theta}{\theta}\right) = \frac{2}{4}\left(1 - \frac{\sin 2.0}{2.0}\right) = 0.273\ ft$$

Applying the Manning equation for flow rate yields

$$Q = \frac{K_m}{n} AR^{2/3} S_o^{\ 1/2} = \frac{1.49}{0.013}(0.545)(0.273)^{2/3}(0.004)^{1/2} = \underline{\underline{1.66\ cfs}}$$

Problem 3.26 Uniform flow (Manning)

Using the Manning equation, compute the normal depth in a 12-m wide rectangular channel carrying 178 m^3/s. The channel is constructed of concrete and is laid on a slope of 0.0028 ft/ft.

Solution

Assuming $n = 0.013$,

$$Q = \frac{K_m}{n} AR^{2/3} S_o^{\ 1/2} = \frac{1.0}{0.013}(12y)\left(\frac{12y}{12 + 2y}\right)^{2/3}(0.0028)^{1/2} = 178$$

Solving for depth yields $y = \underline{\underline{2.5\ m}}$.

Problem 3.27 Uniform flow (Manning)

If the channel in Problem 3.26 has a trapezoidal cross section with a 12-m bottom width and 3H:1V side slopes, what is the normal depth?

Solution

Using $n = 0.013$,

$$Q = \frac{K_m}{n} AR^{2/3} S_o^{\ 1/2}$$

or

$$Q = \frac{1.0}{0.013}\left[(12 + 3y)y\right]\left(\frac{(12 + 3y)y}{12 + 2y\sqrt{1 + (3)^2}}\right)^{2/3}(0.0028)^{1/2} = 178$$

Solving for depth yields $y = \underline{\underline{1.94\ m}}$.

Problem 3.28 Uniform flow (Manning)

Use the Manning equation to determine the depth of flow and velocity in a 3-ft diameter concrete sewer laid on a slope of 0.01 ft/ft and carrying 30 cfs.

Solution

Using $n = 0.013$,

$$Q = \frac{K_m}{n} A R^{2/3} S_o^{1/2}$$

$$= \frac{1.49}{0.013}\left[\frac{(\theta - \sin\theta)(3)^2}{8}\right]\left[\frac{3}{4}\left(1 - \frac{\sin\theta}{\theta}\right)\right]^{2/3}(0.01)^{1/2} = 30$$

Solution by trial-and-error, numerical analysis, or computational solver yields $\theta = 3.0$ rad. Then $A = 3.22$ ft^2, and depth is evaluated as

$$y = \frac{D}{2}\left[1 - \cos\left(\frac{\theta}{2}\right)\right] = \frac{3}{2}\left[1 - \cos\left(\frac{3}{2}\right)\right] = \underline{\underline{1.4\ ft}}$$

Velocity is expressed as

$$V = \frac{Q}{A} = \frac{30}{3.22} = \underline{\underline{9.32\ fps}}$$

Problem 3.29 Uniform flow (Manning)

Determine the normal depth for the parabolic channel shown in Figure P3-29. The channel conveys a discharge of 2.0 m^3/s and has a bed slope of 0.02 m/m. Assume $n = 0.015$.

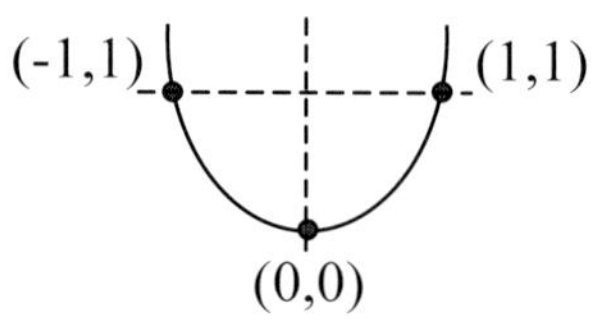

Figure P3-29

Solution

Using the geometric properties in Table 3-1 with $b = 2.0$ m and $Y = 1.0$ m,

$$B = b\sqrt{\frac{y}{Y}} = 2\sqrt{y}$$

$$A = \frac{2}{3}By = \frac{4}{3}y^{3/2}$$

$$x = \frac{4y}{B} = 2\sqrt{y}$$

$$R = \frac{4y}{3}\left[\sqrt{1+x^2} + \left(x^{-1}\right)ln\left(x + \sqrt{1+x^2}\right)\right]^{-1} =$$

$$\frac{4y}{3}\left[\sqrt{1+4y} + \left(\frac{1}{2\sqrt{y}}\right)ln\left(\left(2\sqrt{y}\right) + \sqrt{1+4y}\right)\right]^{-1}$$

Substituting expressions for A and R into the Manning equation gives

$$Q = \frac{K_m}{n} AR^{2/3} S_o^{\;1/2} = \frac{1.0}{0.015} AR^{2/3} (0.02)^{1/2} = 2.0$$

Solving for the normal depth yields y = <u>0.52 m</u>.

Problem 3.30 Flow classification

Determine whether flow from Problem 3.29 is supercritical or subcritical.

Solution

Applying Equation 3-6 with the corresponding geometric properties yields

$$Fr = 1.0 = \frac{BQ^2}{gA^3} = \frac{\left(b\sqrt{\frac{y}{Y}}\right)Q^2}{g\left(\frac{2yb}{3}\sqrt{\frac{y}{Y}}\right)^3} = \frac{3.375Q^2Y}{gb^2y_c^4}$$

Solving for y_c gives

$$y_c = \left(\frac{(3.375)(2.0)^2(1)}{(9.81)(2)^2}\right)^{1/4} = 0.77m$$

Since the normal depth is less than the critical depth, flow is <u>supercritical</u>.

Problem 3.31 Uniform flow (Manning)

A corrugated metal pipe (n = 0.024) has a diameter of 18-in. At its design flow rate, it is intended to flow at a depth of 15 in. At what slope should the pipe be laid so that a minimum self-cleansing velocity of 2 fps is attained.

Solution

From Table 3-1,

$$\theta = 2\cos^{-1}\left(1 - \frac{2y}{D}\right) = 2\cos^{-1}\left(1 - \frac{2(15)}{18}\right) = 4.6\ rad$$

$$A = \frac{(\theta - \sin\theta)D^2}{8} = \frac{(4.6 - \sin 4.6)(18/12)^2}{8} = 1.57\ ft^2$$

$$R = \frac{D}{4}\left(1 - \frac{\sin\theta}{\theta}\right) = \frac{(18/12)}{4}\left(1 - \frac{\sin 4.6}{4.6}\right) = 0.46\ ft$$

Expressing Equation 3-19 in terms of velocity and solving for bed slope yields

$$S_o = \left(\frac{Vn}{K_m R^{2/3}}\right)^2 = \left(\frac{(2.0)(0.024)}{(1.49)(1.57)^{2/3}}\right)^2 = \underline{\underline{5.7 \times 10^{-4}\ ft/ft}}$$

Problem 3.32 Uniform flow (Manning)

Use the Manning equation to determine the minimum diameter of vitrified clay pipe required to carry 45 cfs under gravity conditions over a slope of 0.02 ft/ft. Assume $\nu = 1.2 \times 10^{-5}$ ft^2/s.

Solution

Assuming $n = 0.013$ and uniform flow, Equation 3-20 is written as

$$D = \left(\frac{K_D Q n}{S_o^{1/2}}\right)^{3/8} = \left(\frac{(2.16)(45)(0.013)}{(0.02)^{1/2}}\right)^{3/8} = 2.27\ ft = 27.2\ in$$

This approach assumes that the depth of flow approaches the pipe diameter and that gravity flow prevails. Choosing the next largest commercially available size, a 30-in diameter pipe should be used.

Problem 3.33 Uniform flow (Darcy-Weisbach)

Solve Problem 3.32 using the Darcy-Weisbach equation.

Solution

Assume $k_s = 6 \times 10^{-3}$ ft, and note that for uniform, gravity flow, $S_f = S_o$. Since D is unknown, both the friction factor and velocity are unknown. The problem is solved iteratively beginning with an assumed value for f, say 0.035. Then by Equation 3-21,

$$D = \left(\frac{0.811 f Q^2}{g S_o} \right)^{1/5} = \left(\frac{(0.811)(0.035)(45)^2}{(32.2)(0.02)} \right)^{1/5} = 2.46\ ft$$

The associated velocity and Reynolds number are evaluated as

$$V = \frac{Q}{A} = \frac{45}{\left(\frac{\pi (2.46)^2}{4} \right)} = 9.47\ fps$$

$$Re = \frac{VD}{\upsilon} = \frac{(9.47)(2.46)}{1.2 \times 10^{-5}} = 1.94 \times 10^6$$

Using the Swamee-Jain equation, Equation 3-18, f is checked, as follows:

$$f = \frac{1.325}{\left[ln \left(\frac{k_s}{3.7D} + \frac{5.74}{Re^{0.9}} \right) \right]^2} = \frac{1.325}{\left[ln \left(\frac{6 \times 10^{-3}}{(3.7)(2.46)} + \frac{5.74}{\left(1.94 \times 10^6\right)^{0.9}} \right) \right]^2} = 0.025$$

Since the computed friction factor does not match the assumed value, a second iteration is necessary. Assuming $f = 0.025$, $D = 2.30$ ft, $V = 10.89$ fps, and $Re = 2.0 \times 10^6$. Recomputing f yields 0.025, which matches the assumed value. Therefore, the minimum diameter required is 2.30 ft, or 27.6 in, and a 30-in diameter pipe should be selected.

Comments: An alternative solution method is possible in which D is assumed and checked using the Darcy-Weisbach and Jain equations. Note that the Manning and Darcy-Weisbach equations yield similar results, indicating that a 30-in diameter pipe should be selected.

Problem 3.34 Gravity sewer design

Table P3-34a provides details of the proposed sanitary sewer layout shown in Figure P3-34. Determine the diameter and invert elevations for each length of ductile iron ($n = 0.012$) sewer main. Assume that a peak wastewater flow factor of 2.5 applies, and local standards dictate that a minimum of 4 ft of cover is required and that velocities should range between 2 and 10 fps. Based on uncertainty of flow data, no pipe should flow at a depth more than 75 percent of its diameter.

Solution

Table P3-34b summarizes the computations which lead to specified pipe diameters and invert elevations.

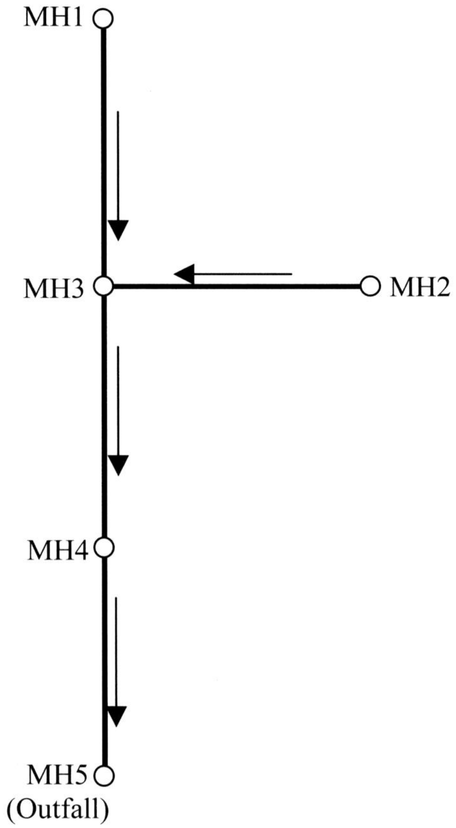

Figure P3-34

Table P3-34a

Sewer	Length (ft)	Ground elevation (ft)		Q_{design} (cfs)
		Upstream manhole	Downstream manhole	
1 - 3	400	98.5	96.1	1.2
2 - 3	400	98.9	96.1	0.8
3 – 4	400	96.1	94.5	3.3
4 – 5	350	94.5	93.1	4.6

Table P3-34b

(1)	(2)	(3)	(4)	(5)	(6)	(7)	(8)	(9)	(10)	(11)	(12)
ID	$S_{surface}$ (ft/ft)	Q_{design} (cfs)	Q_{peak} (cfs)	S_{sewer} (ft/ft)	D (in)	V_{design} (fps)	V_{peak} (fps)	$(y/D)_{design}$ (%)	$(y/D)_{peak}$ (%)	Upper invert (ft)	Lower Invert (ft)
1-3	0.006	1.2	3.0	0.008	**12**	4.0	5.0	40.5	71.7	**93.5**	**90.3**
2-3	0.007	0.8	2.0	0.007	**12**	3.4	4.3	33.8	56.9	**93.9**	**91.1**
3-4	0.004	3.3	8.25	0.007	**18**	4.9	6.1	40.5	71.7	**89.8**	**87.0**
4-5	0.004	4.6	11.5	0.004	**24**	4.3	5.4	37.4	64.0	**86.5**	**85.1**

Specific entries in the Table are as follows:

Column (1): Sewer identification

Column (2): Ground-surface slope is the difference in given ground elevations at corresponding manholes divided by the sewer length

Columns (3) and (4): The design discharge from given data and the peak discharge, computed as the design value multiplied by the peak flow factor

Columns (5) and (6): The sewer slope, or grade, and commercially available diameter are specified values. For a given diameter, the sewer slope is the steeper of the ground slope and the slope required to yield satisfactory velocities and flow depths. If the grade is such that excessive cuts, and thus manhole depths, are required, a new diameter should be selected.

Columns (7) - (10): The velocity and depths for both the design and peak flows are evaluated using the selected given pipe diameter and grade, the corresponding geometric properties from Table 3-1, and the Manning equation (e.g., see Problem 3.28). The velocities should be checked to ensure that minimum and maximum values are satisfied. Likewise, the ratio of depth to diameter is computed and compared to the specified maximum. If criteria are not met, a new diameter and/or slope are specified and the process repeated until a satisfactory design is obtained. Consequently, computer spreadsheets and solvers are convenient for the iterative solution.

Columns (11) and *(12)*: Once diameter and sewer grades are adopted, the invert elevations of the upper and lower ends of a known pipe length are evaluated. The computation begins at the most upstream manhole, where an elevation is specified to satisfy the requirement of a 4-ft cover. Moving downstream, consideration must be given to changing pipe diameters and changing flow directions, if applicable. For example, if the diameter of pipe exiting a manhole is larger than that of the entering pipe, crown elevations should be matched. Doing so prevents full flow from occurring in the smaller pipe as a result of backwater from the larger pipe. Consider, for example, the case of Manhole 3. The lowest invert of entering pipes is 90.3

ft. Thus the invert of the exiting pipe (i.e., Sewer 3 – 4) will be 90.3 – (1.5 – 1), or 89.8 ft. If at any location, minimum cover requirements are not met, a new grade and/or diameter should be selected. For the current analysis, the minimum cover occurs at Manholes 1 and 2 (e.g., 98.9 ft - 93.95 ft – 1 ft pipe diameter = 4.0 ft) and at the downstream end of Sewer 2 – 3. The maximum occurs at Manholes 4 and 5 (e.g., 93.1 ft – 85.1 ft – 2 ft pipe diameter = 6.0 ft).

Comments: The analysis for storm sewers proceeds in a similar manner. The primary difference is that selected diameters affect flow travel times, which indirectly will affect determination of peak discharge (see Problem 2.4). In addition, without the percent-full requirement, (i.e., 75 in this case), Equation 3-20 could be used to directly solve for the minimum required pipe diameter.

Problem 3.35 Gradually varied flow profiles

Sketch water surface profiles for the systems shown in Figure P3-35a. Assume that each channel is sufficiently long to allow complete profiles to develop.

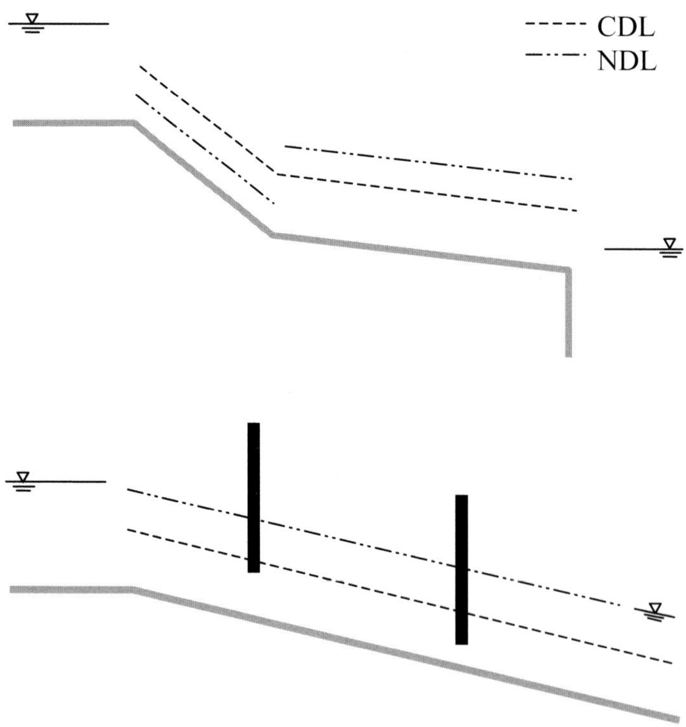

Figure P3-35a

Solution

Once control points (e.g., critical depth at the free overfall or normal depth) have been identified, profiles can be drawn and classified as shown in Figure P3-35b.

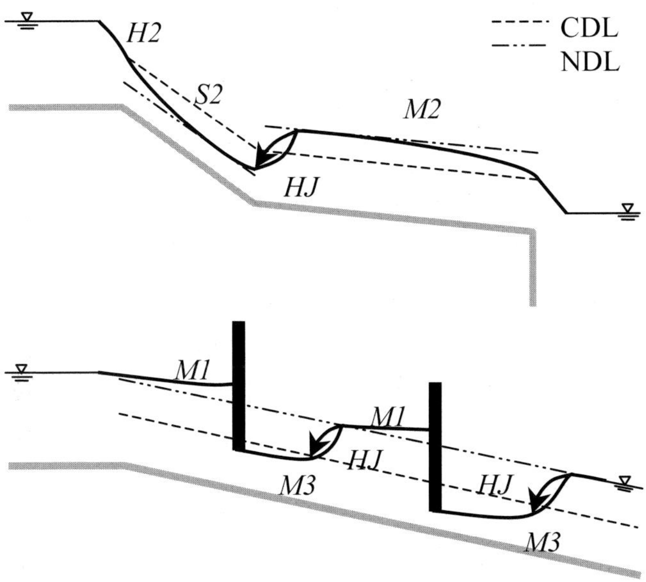

Figure P3-35b

Problem 3.36 Direct step method

A 4-m wide rectangular channel (n = 0.030) carries 8 m^3/s on a slope of 0.005 m/m. If the depth at a particular control section is 3.0 m, how far upstream or downstream is the depth equal to 1.2 m.

Solution

It is first necessary to determine if flow is supercritical or subcritical in order to evaluate the direction of computation. Using previously described methods, the critical and normal depth are computed as 0.74 m and 1.08 m, respectively. Thus, flow is characterized as subcritical and an *M1* water surface profile is being described. Correspondingly, computations will proceed in the upstream direction. Table P3-36 summarizes the direct step computations. Specific entries in the Table are as follows:

Columns (1) and (2): Section identification number and assumed depth. The first entry is the known depth at the control section

Columns (3) – (5): Computed area, hydraulic radius, and velocity for the given channel description and assumed depth of Column 2

Table P3-36

(1) ID	(2) Assumed Depth (m)	(3) A (m^2)	(4) R (m)	(5) V (m/s)	(6) E (m)	(7) S_f (m/m)	(8) $S_{f,avg}$ (m/m)	(9) Δx (m)	(10) Total Distance (m)
1	**3.0**	12.00	1.20	0.67	3.02	3.14×10^{-4}			
							3.62×10^{-4}	- 63.54	**- 63.54**
2	**2.7**	10.80	1.15	0.74	2.73	4.10×10^{-4}			
							4.83×10^{-4}	- 64.78	**- 128.32**
3	**2.4**	9.60	1.09	0.83	2.44	5.57×10^{-4}			
							6.74×10^{-4}	- 66.84	**- 195.15**
4	**2.1**	8.40	1.02	0.95	2.15	7.91×10^{-4}			
							9.92×10^{-4}	- 70.69	**- 265.84**
5	**1.8**	7.20	0.95	1.11	1.86	1.19×10^{-3}			
							1.58×10^{-3}	- 79.62	**- 345.46**
6	**1.5**	6.00	0.86	1.33	1.59	1.97×10^{-3}			
							2.82×10^{-3}	- 114.08	**- 459.53**
7	**1.2**	4.80	0.75	1.67	1.34	3.67×10^{-3}			

Column (6): Specific energy computed from Equation 3-4, using values from Columns 2 and 5

Column (7): Friction slope is computed as

$$S_f = \left(\frac{Vn}{K_m R^{2/3}}\right)^2$$

Column (8): The mean friction slope between the current and previous (i.e., known depth) sections

Column (9): The reach length over which the change in depth occurs is computed using Equation 3-24. Since flow is subcritical, Δx will always be negative, indicating upstream distance.

Column (10): Total distance is the sum of individual reach lengths (i.e., Δx)

Based on the results, a depth of 1.2 m occurs approximately 460 m upstream of the control section.

Problem 3.37 Direct step method

Consider the circular channel and flow rate in Problem 3.28. If critical flow occurs at a particular location, approximately how far upstream or downstream will the depth equal 1.45 ft?

Solution

The normal depth computed from Problem 3.28 was 1.4 ft. The critical depth is computed as 1.8 ft using a similar method to that presented for Problem 3.10. Since $y_c > y_n$, flow is supercritical. The problem thus describes a *S2* profile and computations proceed in the downstream direction (i.e., Δx is positive). The solution is provided in Table P3-37.

Table P3-37

(1)	(2)	(3)	(4)	(5)	(6)	(7)	(8)	(9)	(10)	(11)
ID	Assumed Depth (ft)	θ (rad)	A (ft^2)	R (ft)	V (fps)	E (ft)	S_f (ft/ft)	$S_{f,avg}$ (ft/ft)	Δx (ft)	Total Distance (ft)
1	**1.80**	3.54	4.43	0.83	6.78	2.51	4.46×10^{-3}			
								4.89×10^{-3}	1.11	**1.11**
2	**1.70**	3.41	4.13	0.81	7.26	2.52	5.33×10^{-3}			
								5.91×10^{-3}	7.91	**9.01**
3	**1.60**	3.28	3.83	0.78	7.83	2.55	6.49×10^{-3}			
								7.27×10^{-3}	24.93	**33.94**
4	**1.50**	3.14	3.53	0.75	8.49	2.62	8.05×10^{-3}			
								8.54×10^{-3}	35.27	**69.21**
5	**1.45**	3.08	3.38	0.73	8.86	2.67	9.04×10^{-3}			

Note that the solution method is the same as that of Problem 3.36, except that θ must be computed and different formulae for area and hydraulic radius are required. Based on the Table provided, a depth of 1.45 ft occurs nearly <u>70 ft downstream</u> of the critical flow control section.

Problem 3.38 Direct step method

Consider the parabolic channel and flow rate in Problems 3.29 and 3.30. Over what distance would depth change from 0.3 m to 0.4 m?

Solution

From Problem 3.29, flow is supercritical. The problem describes an *S1* profile and computations will proceed in the downstream direction. At the upstream section, $y = 0.3$ m, and

$$B = 2\sqrt{0.3} = 1.10\ m$$

$$A = \frac{4}{3}(0.3)^{3/2} = 0.22\ m^2$$

$$x = 2\sqrt{0.3} = 1.10$$

$$R = \frac{4(0.3)}{3}\left[\sqrt{1+4(0.3)} + \left(\frac{1}{2\sqrt{0.3}}\right) ln\left(\left(2\sqrt{0.3}\right) + \sqrt{1+4(0.3)}\right)\right]^{-1} = 0.17\ m$$

$$V = \frac{Q}{A} = \frac{2.0}{0.22} = 9.09\frac{m}{s}$$

$$E = y + \frac{V^2}{2g} = 0.3 + \frac{(9.09)^2}{2(9.81)} = 4.51\ m$$

Using a similar approach, the hydraulic radius, velocity, and specific energy at the downstream section, where y = 0.4 m, are equal to 0.22 m, 5.88 m/s and 2.16 m, respectively. The average velocity and hydraulic radius between the two sections is, thus, 0.195 m and 7.49 m/s, respectively. The friction slope is evaluated as

$$S_f = \left(\frac{(7.49)(0.015)}{1.0(0.195)^{2/3}}\right)^2 = 0.112\ m$$

The distance between the two sections is expressed using Equation 3-24 as follows:

$$x_{i+1} - x_i = \frac{E_{i+1} - E_i}{S_o - S_f} = \frac{2.16 - 4.51}{0.02 - 0.112} = \underline{\underline{25.5\ m}}$$

Problem 3.39 Standard step method

Consider the channel and control section described in Problem 3.36. What is the depth at a section 500 m upstream of the given control depth?

Solution

Flow is subcritical (see Problem 3.36), and thus the standard step method can be applied to determine the upstream depth. Table P3-39 summarizes the gradually varied flow computation.

Table P3-39

(1)	(2)	(3)	(4)	(5)	(6)	(7)	(8)	(9)	(10)
Total Dist. (m)	Reach Length (m)	Assumed Depth (m)	A (m^2)	V (m/s)	E (m)	R (m)	S_f (m/m)	$S_{f,avg}$ (m/m)	E (m)
0	0	**3.00**	12.00	0.67	**3.02**	1.20	3.14×10^{-4}	-	-
100	100	**2.53**	10.12	0.79	**2.56**	1.12	4.85×10^{-4}	3.99×10^{-4}	**2.56**
200	100	**2.08**	8.32	0.96	**2.13**	1.02	8.11×10^{-4}	6.48×10^{-4}	**2.13**
300	100	**1.67**	6.67	1.20	**1.74**	0.91	1.47×10^{-3}	1.14×10^{-3}	**1.74**
400	100	**1.34**	5.34	1.50	**1.45**	0.80	2.71×10^{-3}	2.09×10^{-3}	**1.45**
500	100	**1.14**	4.56	1.75	**1.30**	0.73	4.23×10^{-3}	3.47×10^{-3}	**1.30**

Specific entries are as follows:

Columns (1) and (2): Total distance is the sum of assumed reach lengths in Column 2

Column (3): The first entry is the known depth at the control section, while remaining entries are assumed values based on results of Columns 6 and 10

Columns (4) and (5): Computed area and velocity for the given channel description and assumed depth

Column (6): Specific energy computed from Equation 3-4, using values from Columns 3 and 5

Column (7): Hydraulic radius for the assumed depth

Column (8): Friction slope is computed as

$$S_f = \left(\frac{Vn}{K_m R^{2/3}} \right)^2$$

Column (9): The mean friction slope between the current and previous (i.e., known depth) sections

Column (10): The specific energy computed via Equation 3-26. For the subcritical case, this is expressed as

$$E_i = E_{i+1} - S_o \Delta x + S_{f,avg} \Delta x$$

The objective is to assume a depth which satisfies Equation 3-26, or alternatively stated, which causes the values in Columns 6 and 10 to be equal. Table P3-39 shows the final depths and indicates that y is approximately <u>1.14</u> m at a distance of 500 m upstream of the control section.

Problem 3.40 Standard step method

A concrete trapezoidal channel carries 15 m^3/s over a slope of 0.05 m/m. The channel has a bottom width of 5 m and side slopes of 2H:1V. What is the depth 100 m downstream of a control section where critical flow prevails?

Solution

Assuming $n = 0.013$, the critical and normal depths are evaluated as 0.86 m and 0.34 m, respectively. Flow is supercritical, and computations proceed in the downstream direction. The analysis proceeds similar to that of Problem 3.39. The primary differences are the use of different formulae for area and hydraulic radius and the form of Equation 3-26. For supercritical conditions, specific energy is evaluated as

$$E_i = E_{i+1} + S_o \Delta x - S_{f,avg} \Delta x$$

Table P3-40 summarizes the gradually varied flow computation and indicates that depth at a distance of 100 m downstream is equal to 0.35 m.

Problem 3.41 Hydraulic gradient evaluation and form losses

A 3.0-ft diameter concrete ($n = 0.013$) storm sewer carries a discharge of 30 cfs on a slope of 0.004 ft/ft. The 300-ft long pipe includes four 30° bends, located at distances of 50 ft, 100 ft, 150 ft, and 200 ft from its downstream end. If the tailwater elevation is 2 ft, relative to the pipe's downstream invert, determine the HGL elevation at the upstream end.

Solution

Normal and critical depths are computed as 1.87 ft and 1.77 ft, respectively. Since the tailwater elevation lies below the crown elevation and since flow is subcritical, computations can begin downstream. Also, since the tailwater is greater than the critical depth, the control elevation is taken as the greater of the tailwater or $(D + y_c)/2$. In this case, the latter indicates that the control elevation should be 2.39 ft. The solution is similar to that of Problem 3.39 and is summarized in Table P3-41. Note, however, that θ must be computed and that different formulae for area and hydraulic radius are required.

In addition, Column 10 is added and includes form losses created by the bends. The loss associated with each bend is computed as

$$h_l = 0.25\sqrt{\frac{30}{90}}\left(\frac{(V)^2}{2(32.2)}\right)$$

where V is the velocity at the bend location. The specific energy in Column 11 is computed as

$$E_i = E_{i+1} - S_o \Delta x + S_{f,avg} \Delta x + h_l$$

The objective is to assume a depth which satisfies the equation (i.e., causing the values in Columns 6 and 11 to be equal). The results indicate that an *M1* profile exists and that the upstream HGL, relative to the pipe invert at that location, is equivalent to 1.872 ft. Relative to the downstream invert, the upstream HGL elevation is 1.872 ft + 0.004 ft/ft × 300 ft, or 3.07 ft.

Table P3-40

(1)	(2)	(3)	(4)	(5)	(6)	(7)	(8)	(9)	(10)
Total Dist. (m)	Reach Length (m)	Assumed Depth (m)	A (m^2)	V (m/s)	E (m)	R (m)	S_f (m/m)	$S_{f,avg}$ (m/m)	E (m)
0	0	**0.86**	5.78	2.60	**1.20**	0.653	2.01×10^{-3}	-	-
25	25	**0.44**	2.58	5.81	**2.16**	0.371	2.15×10^{-2}	1.17×10^{-2}	**2.16**
50	25	**0.38**	2.22	6.77	**2.72**	0.330	3.40×10^{-2}	2.77×10^{-2}	**2.72**
75	25	**0.36**	2.08	7.22	**3.02**	0.314	4.13×10^{-2}	3.76×10^{-2}	**3.02**
100	25	**0.35**	2.01	7.45	**3.19**	0.306	4.54×10^{-2}	4.33×10^{-2}	**3.19**

Table P3-41

(1)	(2)	(3)	(4)	(5)	(6)	(7)	(8)	(9)	(10)	(11)
Total Dist. (ft)	Reach Length (ft)	Assumed Depth (ft)	A (ft^2)	V (fps)	E (ft)	R (ft)	S_f (ft/ft)	$S_{f,avg}$ (ft/ft)	h_l (ft)	E (ft)
0	0	**2.39**	6.04	4.97	**2.77**	0.912	2.12×10^{-3}	-		**2.77**
50	50	**2.34**	5.92	5.07	**2.74**	0.911	2.22×10^{-3}	2.17×10^{-3}	0.058	**2.74**
100	50	**2.21**	5.58	5.38	**2.66**	0.902	2.53×10^{-3}	2.37×10^{-3}	0.065	**2.66**
150	50	**2.08**	5.25	5.72	**2.59**	0.887	2.92×10^{-3}	2.72×10^{-3}	0.073	**2.59**
200	50	**1.98**	4.96	6.05	**2.55**	0.870	3.36×10^{-3}	3.14×10^{-3}	0.082	**2.55**
250	50	**1.90**	4.74	6.33	**2.53**	0.856	3.76×10^{-3}	3.56×10^{-3}	-	**2.53**
300	50	**1.872**	4.64	6.47	**2.52**	0.849	3.96×10^{-3}	3.86×10^{-3}	-	**2.52**

Problem 3.42 Hydraulic jump

A 10-ft wide rectangular channel (n = 0.015) carries 400 cfs at a normal depth of 3 ft. An obstruction causes the flow depth just upstream of the obstruction to be 8 ft. Will a hydraulic jump form upstream of the obstruction? If so, what is the associated energy loss?

Solution

The critical depth can be evaluated by

$$y_c = \sqrt[3]{\frac{q^2}{g}} = \sqrt[3]{\frac{\left(400/10\right)^2}{32.2}} = 3.68\ ft$$

Note that $y_n < y_c < 8$ ft, and thus supercritical flow will prevail further upstream and a hydraulic jump will form. Assuming that the depth immediately before the jump is the normal depth, the sequent depth can be found by Equation 3-30 as follows:

$$y_2 = \frac{y_1}{2}\left(-1+\sqrt{1+8Fr_1^2}\right) = \frac{3}{2}\left(-1+\sqrt{1+8\left(\frac{(400/30)}{\sqrt{(32.2)(3)}}\right)^2}\right) = 4.45\ ft$$

Thus an *M1* profile is created upstream of the obstruction. If needed, the direct step method could be used to determine the distance upstream from the obstruction to the jump (i.e., distance between 4.45 ft and 8 ft depths). The energy loss through the jump is evaluated by computing specific energy for the sequent depths.

$$E_1 = y_1 + \frac{Q^2}{2gA_1^2} = 3 + \frac{(400)^2}{2(32.2)(10\times 3)^2} = 5.76\ ft$$

$$E_2 = y_2 + \frac{Q^2}{2gA_2^2} = 4.45 + \frac{(400)^2}{2(32.2)(10\times 4.45)^2} = 5.70\ ft$$

Thus, the energy loss is $E_1 - E_2$, or 0.06 ft.

Problem 3.43 Hydraulic jump

Consider a hydraulic jump in a 3-ft diameter sewer carrying 6 cfs. If the upstream depth is 0.6 ft, evaluate the downstream sequent depth.

Solution

The critical depth is approximated from Equation 3-8 as

$$y_c = \left(\frac{1.01}{D^{0.26}}\right)\left(\frac{Q^2}{g}\right)^{0.25} = \left(\frac{1.01}{(3)^{0.26}}\right)\left(\frac{(6)^2}{32.2}\right)^{0.25} = 0.78\ ft$$

and the corresponding value of θ is

$$\theta_c = 2\cos^{-1}\left(1-\frac{2y}{D}\right) = 2\cos^{-1}\left(1-\frac{2(0.78)}{3}\right) = 2.14\ rad$$

For the given supercritical depth,

$$\theta = 2\cos^{-1}\left(1-\frac{2y}{D}\right) = 2\cos^{-1}\left(1-\frac{2(0.6)}{3}\right) = 1.85\ rad$$

The upstream specific momentum is evaluated using the formula given in Table 3-4 as follows

$$M_1 = \frac{D^3}{24}\left[3\sin\left(\frac{\theta}{2}\right) - \sin^3\left(\frac{\theta}{2}\right) - \frac{3\theta}{2}\cos\left(\frac{\theta}{2}\right)\right] + \frac{Q^2}{gD^2(\theta-\sin\theta)/8}$$

or

$$\frac{(3)^3}{24}\left[3\sin\left(\frac{1.85}{2}\right) - \sin^3\left(\frac{1.85}{2}\right) - \frac{3(1.85)}{2}\cos\left(\frac{1.85}{2}\right)\right] + \frac{(6)^2}{(32.2)(3)^2(1.85-\sin 1.85)/8} = 1.36\ ft^3$$

Then from Equation 3-28 (i.e., $M_1 = M_2$), the downstream specific momentum is

$$1.36 = \frac{(3)^3}{24}\left[3\sin\left(\frac{\theta}{2}\right) - \sin^3\left(\frac{\theta}{2}\right) - \frac{3\theta}{2}\cos\left(\frac{\theta}{2}\right)\right] + \frac{(6)^2}{(32.2)(3)^2(\theta-\sin\theta)/8}$$

Solving for θ, and limiting the search to $\theta > \theta_c$, yields 2.42 rad, or

$$y = \frac{D}{2}\left[1-\cos\left(\frac{\theta}{2}\right)\right] = \frac{3}{2}\left[1-\cos\left(\frac{2.42}{2}\right)\right] = \underline{\underline{0.97\ ft}}$$

Problem 3.44 Hydraulic jump

A parabolic channel has a channel top width of 9 m when the depth is 2 m. If the depth downstream of a hydraulic jump is 1.6 m for a discharge of 8.5 m^3/s, what is the upstream sequent depth?

Solution

Using the formula derived in Problem 3.30, the critical depth is evaluated as

$$y_c = \left(\frac{(3.375)(Q)^2(Y)}{(g)(b)^2}\right)^{1/4} = \left(\frac{(3.375)(8.5)^2(2)}{(9.81)(9)^2}\right)^{1/4} = 0.89\,ft$$

Using information from Table 3-4, the downstream specific momentum is evaluated as

$$M_2 = \frac{4}{15}\frac{by^{5/2}}{\sqrt{Y}} + \frac{1.5Q^2\sqrt{Y}}{gby^{3/2}} = \frac{4}{15}\frac{(9)(1.6)^{5/2}}{\sqrt{2}} + \frac{1.5(8.5)^2\sqrt{2}}{(9.81)(9)(1.6)^{3/2}} = 6.35\,ft^3$$

Knowing that $M_1 = M_2$, the upstream specific momentum is expressed as

$$M_1 = \frac{4}{15}\frac{by^{5/2}}{\sqrt{Y}} + \frac{1.5Q^2\sqrt{Y}}{gby^{3/2}} = \frac{4}{15}\frac{(9)y^{5/2}}{\sqrt{2}} + \frac{1.5(8.5)^2\sqrt{2}}{(9.81)(9)y^{3/2}} = 6.35\,ft^3$$

Solving for the supercritical root yields $y = 0.43$ m.

REFERENCES CITED

American Iron and Steel Institute (AISI), *Modern Sewer Design*, 2nd ed., AISI, Washington, DC, 1990.

American Society of Civil Engineers (ASCE), *Gravity Sanitary Sewer Design and Construction*, ASCE Manuals and Reports on Engineering Practice No. 60 and Water Pollut. Control Fed. Manual of Practice FD-5, New York, NY, 1982.

Bombardelli, F.A., and M.H. Garcia, "Hydraulic Design of Large-Diameter Pipes," *J. of Hydraulic Engrg.*, ASCE, vol. 129, no. 11, 839-846, 2003.

Brown, S.A., S.M. Stein, and J.C. Warner, *Urban Drainage Design Manual*, Hydraulic Engineering Circular No. 22, FHWA-SA-96-078, Federal Highway Administration, U.S. Department of Transportation, Washington, DC, 1996.

Chow, V.T., *Open Channel Hydraulics*, McGraw-Hill, New York, NY, 1959.

Crowe, C.T., D.E. Elger, and J.A. Roberson, *Engineering Fluid Mechanics*, 7th ed., Wiley, New York, NY, 2001.

French, R.H., *Open-Channel Hydraulics*, McGraw-Hill, New York, NY, 1985.

Henderson, F.M., *Open Channel Flow*, Macmillan, New York, NY, 1966.

Jain, A.K., D.M. Mohan, and P. Khanna, "Modified Hazen-Williams Formula," *J. Environ. Engrg. Div.*, ASCE, vol. 104, no. 1, 137-146, 1978.

Lindeburg, M.R., *Civil Engineering Reference Manual*, 6th ed., Professional Publications, Belmont, CA, 1992.

Sangster, W.M., H.M. Wood, E.T. Smerdon, and H.G. Bossy, "Pressure Changes at Storm Drain Junctions," Engineering Series Bulletin, No. 41, Engineering Experiment Station, University of Missouri, Columbia, MO, 1958.

Sturm, T.W., *Open Channel Hydraulics*, McGraw-Hill, New York, NY, 2001.

Swamee, P.K., and A.K. Jain, "Explicit Equations for Pipe-Flow Problems," *J. of Hydraulic Div.*, ASCE, vol. 102, no. 5, 657-664, 1976.

U.S. Bureau of Reclamation (USBR), "Research Studies on Stilling Basings, Energy Dissipators, and Associated Appurtenances," Hydraulic Laboratory Report No. HYD-399, 1955.

Water Environment Federation (WEF), *Design of Wastewater and Stormwater Pumping Stations*, WEF Manual of Practice FD-4, Alexandria, VA, 1993.

CHAPTER FOUR
Unsteady Flow Hydraulics

4.1 INTRODUCTION

Unsteady flow occurs when velocity changes over time at a fixed location. Most unsteady flows are also categorized as nonuniform, either rapidly or gradually varied, implying that velocity changes with location along the sewer. For the case of rapidly varied flows, steep wave fronts and discontinuities may occur. Such conditions are referred to as shocks or bores. Subsequently, the analysis of these and other unsteady flow problems requires solution to an appropriate set of governing equations in order to describe the spatial and temporal distribution of flow velocity and depth.

4.2 GOVERNING EQUATIONS

Sewer flows can be described using conservation laws for mass, energy, and momentum. Since flow characteristics can be defined by a combination of two parameters, such as velocity and depth, only two governing equations are generally required. Assuming the velocity is uniform over a given cross section, the momentum and energy equations are equivalent (Cunge et al., 1980). For cases in which depth and velocity are discontinuous (e.g., hydraulic jump or storm surge), the momentum equation should be used since it is not necessary to know the amount of energy lost at the discontinuity. The complete differential forms of the momentum and continuity equations are together commonly referred to as the St. Venant (i.e., Saint-Venant) equations and can be used to describe unsteady flow in sewers.

The conservation (i.e., discharge-area) form of the continuity equation can be expressed as

$$\frac{\partial Q}{\partial x} + \frac{\partial A}{\partial t} = 0 \qquad (4\text{-}1)$$

where Q is volumetric flow rate; x is longitudinal distance; A is effective flow area; and t is time. The momentum equation can be written as

$$\frac{1}{A}\frac{\partial Q}{\partial t} + \frac{1}{A}\frac{\partial (Q^2/A)}{\partial x} + g\frac{\partial y}{\partial x} - g(S_o - S_f) = 0 \qquad (4\text{-}2)$$

where g is acceleration due to gravity; y is flow depth; S_o is the longitudinal channel slope; and S_f is the friction slope, or slope of the EGL. The corresponding nonconservation forms (i.e., velocity-depth) of the St. Venant equations are

$$V\frac{\partial y}{\partial x} + D\frac{\partial V}{\partial x} + \frac{\partial y}{\partial t} = 0 \tag{4-3}$$

and

$$\frac{\partial V}{\partial t} + V\frac{\partial V}{\partial x} + g\frac{\partial y}{\partial x} - g\left(S_o - S_f\right) = 0 \tag{4-4}$$

where D is hydraulic depth, or effective area, A, divided by top width, B. Several assumptions inherent to the derivation of the St. Venant equations as written are that lateral inflow or outflow (e.g., infiltration) is negligible; the pressure distribution remains hydrostatic; the channel slope is small; the velocity distribution over a cross section is uniform; the channel is prismatic; and that energy losses (i.e., S_f) are the same as for the steady flow case and can therefore be simulated using a relationship such as the Manning equation.

It is worth noting the significance of each term in Equation 4-2 or 4-4. For instance, the first and second terms in the momentum equation are the local and convective acceleration terms, respectively. Those terms remaining are pressure, gravity, and friction force components. A variety of flow models can be created by combining continuity with alternative forms of the momentum equation. For example, the kinematic wave model neglects local and convective acceleration terms, as well as pressure force terms, thus yielding

$$S_o - S_f = 0 \tag{4-5}$$

which is directly applicable to steady, uniform flows. Likewise, the diffusion wave model that assumes steady, non-uniform flow is developed by neglecting only acceleration components, and the dynamic wave model considers the full momentum equation as written (i.e., unsteady and nonuniform).

4.3 SOLUTION TECHNIQUES

Only in a limited number of simplified cases can the St. Venant equations be solved analytically. They represent a set of nonlinear, partial differential equations for which a closed-form solution is not readily available. Thus, for most applications, numerical techniques are required. Some of the more common techniques include the method of characteristics, finite difference techniques, the kinematic wave model, and the Muskingum-Cunge method. Each of these methods is briefly explained, primarily with respect to unsteady flow with a free surface.

4.3.1 Method of Characteristics

The method of characteristics is based on the premise that the continuity and momentum equations for a prismatic channel can be rewritten for total derivatives, D, as

$$\frac{\mathrm{D}V}{\mathrm{D}t} = \frac{\partial V}{\partial t} + \frac{\partial V}{\partial x}\frac{dx}{dt} \tag{4-6}$$

and

$$\frac{\mathrm{D}y}{\mathrm{D}t} = \frac{\partial y}{\partial t} + \frac{\partial y}{\partial x}\frac{dx}{dt} \tag{4-7}$$

in order to derive the following equivalent set of compatibility equations:

$$\frac{d(V + 2c)}{dt} = g\left(S_o - S_f\right) \tag{4-8}$$

and

$$\frac{d(V - 2c)}{dt} = g\left(S_o - S_f\right) \tag{4-9}$$

where c is the dynamic wave celerity, or $\sqrt{g\,A/B}$, which is the speed at which a wave travels along the channel. Equation 4-8 is valid if

$$\frac{dx}{dt} = V + c \tag{4-10}$$

while Equation 4-9 is valid if

$$\frac{dx}{dt} = V - c \tag{4-11}$$

Equation 4-10 and 4-11 represent curves if plotted in an *x-t* coordinate system and are referred to as positive (i.e., C^+) and negative (i.e., C^-) characteristics. By multiplying Equations 4-8 through 4-11 by dt, integrating, and applying the trapezoidal rule to evaluate the resulting integrals, the following four nonlinear algebraic equations result:

$$x_P = x_A + (t_P - t_A)\left[\frac{V_P + c_P + V_A + c_A}{2}\right] \tag{4-12}$$

$$V_P = V_A + 2(c_A - c_P) + \frac{g}{2}(t_P - t_A)\left[(S_o - S_f)_P + (S_o - S_f)_A\right] \tag{4-13}$$

$$x_P = x_B + (t_P - t_B)\left[\frac{V_P - c_P + V_B - c_B}{2}\right] \tag{4-14}$$

$$V_P = V_B + 2(c_P - c_B) + \frac{g}{2}(t_P - t_B)\left[(S_o - S_f)_P + (S_o - S_f)_B\right] \tag{4-15}$$

where subscripts A, P, and B refer to x-t coordinates, as shown in Figure 4-1.

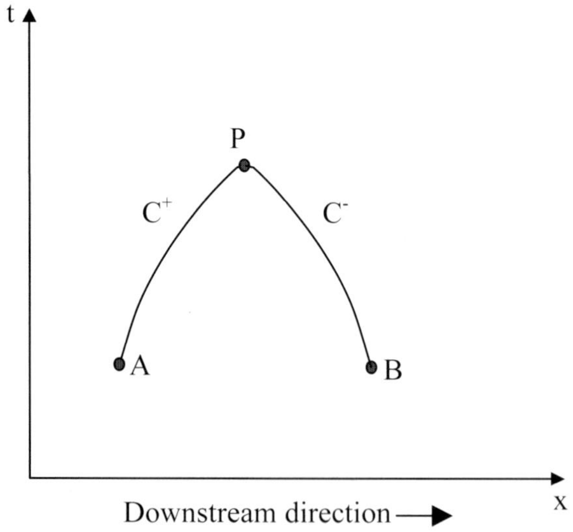

Figure 4-1: Characteristics in the *x-t* coordinate system

Thus, if conditions are known at A and B, Equations 4-12 through 4-15 can be solved simultaneously for the four unknowns (i.e., V_P, y_P, x_P, and t_P). Similar successive relationships can then be written to evaluate the entire spatial and temporal variation of depth and velocity. Note that the locations of computational nodes (i.e., P) are based upon values of previously-computed velocities and depths and are determined by the intersection of the characteristics. Thus, time and space increments are not specified in the approach described here. If conditions are sought at specific locations or times, they must be evaluated by interpolation of known values at neighboring points.

The exceptions to solving only the compatibility and characteristic equations occur at the upstream and downstream limits of the physical system and at the simulation start time (i.e., $t = 0$). At these locations and time, boundary and initial conditions must be specified. These conditions should be independent of the governing equations used in the solution and may consist of a specified depth, velocity, or relationship between V and y. For subcritical flows, one condition is needed at the upstream boundary, and one is needed at the downstream boundary. For supercritical flows, two conditions must be specified at the upstream boundary, and none are needed downstream. For additional information regarding boundary conditions, as well as other detailed aspects of the method of characteristics, see Sturm (2001) or Chaudhry (1993).

4.3.2 Finite Difference Techniques

Finite difference methods involve approximation of the differential form of the continuity and momentum equation using difference equations, which in turn form a set of solvable algebraic equations at discrete points in time and space. The methods can be categorized as either explicit or implicit schemes. Considering the time-space grid shown in Figure 4-2 and the functional derivative $\partial f / \partial x$ (e.g., $\partial V / \partial x$ or $\partial Q / \partial x$), explicit methods are those that approximate the derivative in terms of variables at the known time level (i.e., k), or

$$\frac{\partial f}{\partial x} = \frac{f_i^k - f_{i-1}^k}{\Delta x} \tag{4-16}$$

Since this equation is written in terms of known quantities, the transformed set of algebraic equations can be solved directly. Equation 4-16 is also referred to as a first-order, backward approximation since variables at the known and previous distance levels (i.e., i and $i-1$) are used. Forward approximations rely upon $i+1$ and i, while central difference approximations are based on locations $i+1$ and $i-1$. The latter requires using $2\Delta x$ in the denominator and is thus second-order accurate. Generally speaking, explicit methods are more accurate and often simple to implement, but for reasons of stability (i.e., whether error grows or decays as the solution progresses), they require smaller computational time intervals and thus potentially greater computational time.

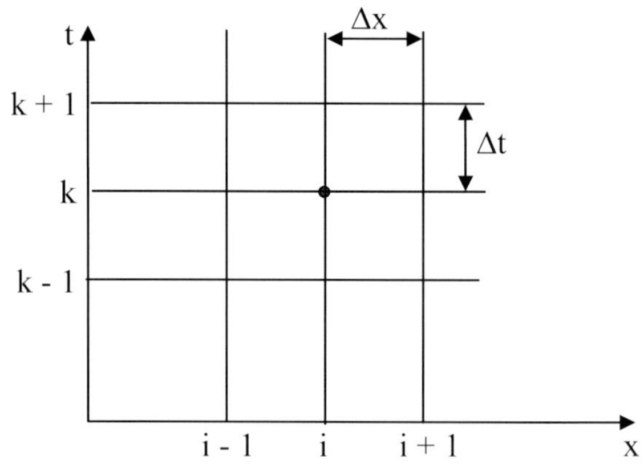

Figure 4-2: Finite difference grid

In contrast to explicit techniques, implicit methods rely upon variables at the unknown time $k+1$ and therefore require the simultaneous solution of a set of finite difference algebraic relationships. The implicit counterpart to Equation 4-16 could be written as

$$\frac{\partial f}{\partial x} = \frac{f_i^{k+1} - f_{i-1}^{k+1}}{\Delta x} \tag{4-17}$$

Implicit methods are inherently more stable and less sensitive to specification of time and space increments.

Although numerous finite difference methods exist, the focus here is on a weighted, four-point, implicit method, also known as the Preissmann scheme. In this case, space and time derivatives are approximated as

$$\frac{\partial f}{\partial x} = \frac{\theta\left(f_{i+1}^{k+1} - f_i^{k+1}\right) + (1-\theta)\left(f_{i+1}^{k} - f_i^{k}\right)}{\Delta x} \tag{4-18}$$

and

$$\frac{\partial f}{\partial t} = \frac{\left(f_i^{k+1} + f_{i+1}^{k+1}\right) - \left(f_i^{k} + f_{i+1}^{k}\right)}{2\Delta t} \tag{4-19}$$

and where nonderivative terms, f, can be evaluated as

$$f = \theta\frac{\left(f_{i+1}^{k+1} + f_i^{k+1}\right)}{2} + (1-\theta)\frac{\left(f_{i+1}^{k} + f_i^{k}\right)}{2} \tag{4-20}$$

where θ is a weighting coefficient. Application of this approximation scheme to the continuity and momentum equations yields

$$\frac{\Delta x}{2\Delta t}\left(A_i^{k+1} + A_{i+1}^{k+1} - A_i^k - A_{i+1}^k\right) + \theta\left[(VA)_{i+1}^{k+1} - (VA)_i^{k+1}\right] + (1-\theta)\left[(VA)_{i+1}^k - (VA)_i^k\right] = 0 \tag{4-21}$$

and

$$\frac{\Delta x}{2\Delta t}\left[(VA)_i^{k+1} + (VA)_{i+1}^{k+1} - (VA)_i^k - (VA)_{i+1}^k\right] + \theta\left\{\left(V^2A\right)_{i+1}^{k+1} - \left(V^2A\right)_i^{k+1} + g\overline{A}^{k+1}\left[y_{i+1}^{k+1} - y_i^{k+1} - \Delta x\left(\overline{S_o - S_f}\right)^{k+1}\right]\right\} + (1-\theta)\left\{\left(V^2A\right)_{i+1}^k - \left(V^2A\right)_i^k + g\overline{A}^k\left[y_{i+1}^k - y_i^k - \Delta x\left(\overline{S_o - S_f}\right)^k\right]\right\} = 0 \tag{4-22}$$

where

$$\overline{A} = \frac{A_{i+1} + A_i}{2} \tag{4-23}$$

and

$$\overline{S_o - S_f} = \frac{\left(S_o - S_f\right)_{i+1} + \left(S_o - S_f\right)_i}{2} \tag{4-24}$$

Selection of various values for the weighting coefficient can make the scheme fully explicit (i.e., $\theta = 0$) or fully implicit (i.e., $\theta = 1$). The method is only stable, however, when $0.55 < \theta \leq 1$, and typical values range from 0.6 to 0.7 (Chaudhry, 1993). Equations 4-21 and 4-22 are a function of four unknowns (i.e., V_i^{k+1}, y_i^{k+1}, V_{i+1}^{k+1}, and y_{i+1}^{k+1}) and can be written for each interior point in the *x-t* grid. Along with appropriate boundary conditions, this set of nonlinear algebraic equations can be solved simultaneously for the spatial and temporal variation of velocity and depth. Numerical techniques such as the Gauss-Jordan and Newton-Raphson methods often prove useful in solving this system of equations.

The implicit scheme is generally stable for any selected value of Δt and Δx, assuming that the weighting coefficient is properly selected. For accuracy, however, the Courant number, C_n, should be close to unity. The Courant number represents a ratio of actual dynamic wave celerity to that computed numerically, or

$$C_n = \frac{|V| \pm c}{\Delta x / \Delta t} \tag{4-25}$$

4.3.3 Kinematic Wave Models

Kinematic wave theory relies on the continuity equation and a simplified form of the momentum equation to derive solutions. From the perspective of momentum conservation, flow is assumed to be steady and uniform from one time increment to the next (i.e., $S_o = S_f$); the unsteady aspect, however, is maintained within the continuity equation. The implication is that simulated kinematic waves will not appreciably accelerate and can only flow in the downstream direction. Thus the wave will be observed as a relatively uniform rise and fall in the water surface over a long period.

For uniform flow, the momentum equation can be expressed in the general form

$$Q = \alpha_k A^m \tag{4-26}$$

For example, the Manning equation (see Chapter 3) can be similarly written as

$$Q = \left(\frac{K_m S_o^{1/2}}{n P^{2/3}} \right) A^{5/3} \tag{4-27}$$

The kinematic wave celerity, c_k, defined as dQ/dA for a unique relationship between area and discharge, can thus be expressed as

$$c_k = m \alpha_k A^{m-1} = mV \tag{4-28}$$

where V is flow velocity based on a reference discharge and is computed using the Manning or similar equation. Applying the chain rule and substituting the result into the continuity equation yields

$$\frac{\partial Q}{\partial t} + c_k \frac{\partial Q}{\partial x} = 0 \tag{4-29}$$

which is the governing relationship for movement of a kinematic wave. Ponce (1989) suggested that the results from Equation 4-29 are valid if

$$\frac{T_r V S_o}{y} > 85 \tag{4-30}$$

where T_r is the time associated with the rising limb of the associated hydrograph

Generally, c_k would be expected to vary with depth and, thus, discharge. The assumption of a constant celerity, however, is sometimes used, whereby c_k corresponds to a particular reference discharge (e.g., peak, average, mid-range value). In this case, the kinematic wave equation can be solved analytically. The solution would represent straightforward translation of a hydrograph, without a change in shape or attenuation in peak discharge. For example, if a particular discharge enters a channel at time t_o, then the time at which that discharge appears at a distance L downstream can be expressed as

$$t = t_o + \frac{L}{c_k} \tag{4-31}$$

If c_k is not constant, the shape and height of the kinematic wave, and thus the hydrograph, changes as it propagates downstream. The implication is that Equation 4-29 requires a numerical solution. Using an implicit, four-point, forward finite difference scheme, the kinematic wave equation becomes

$$\begin{aligned} &\frac{\Delta x}{2\Delta t}\left(Q_i^{k+1} + Q_{i+1}^{k+1} - Q_i^k - Q_{i+1}^k\right) \\ &\qquad + \overline{c_k}\left(Q_{i+1}^{k+1} - Q_i^{k+1}\right) = 0 \end{aligned} \tag{4-32}$$

which can be solved for Q_{i+1}^{k+1} using an iterative technique such as the Newton-Raphson method.

For a nonlinear model, the value of $\overline{c_k}$ in Equation 4-32 could be interpreted in terms of unknown conditions (e.g., V, Q, or A) at time $k+1$, or

$$\overline{c_k} = \frac{c_{k,i}^{k+1} + c_{k,i+1}^{k+1}}{2} \tag{4-33}$$

For a linear model, $\overline{c_k}$ could be expressed by averaging known values diagonally across each time and space interval, or

$$\overline{c_k} = \frac{c_{k,i}^{k+1} + c_{k,i+1}^{k}}{2} \tag{4-34}$$

In addition, recall that c_k is defined as dQ/dA if discharge is solely a function of flow area, or is approximated as so. This amounts to a simple, convenient relationship in most cases. For example, using the Manning equation for a wide, rectangular channel, m from Equation 4-26 assumes a value of 5/3 since wetted perimeter is independent of flow area. Thus, from Equation 4-28, celerity could be evaluated as

$$c_k = \frac{5}{3}V = \frac{5}{3}\frac{K_m}{n}\left(\frac{A}{P}\right)^{2/3} S_o^{1/2} \tag{4-35}$$

If area is not independent of wetted perimeter, and thus depth, c_k should be defined as $\partial Q/\partial A$. For example, if the Manning equation is used to evaluate discharge in a partially full circular sewer of diameter D, then

$$c_k = \frac{0.132K_m}{n} S_o^{1/2} D^{2/3}\left(1 - \frac{\sin\theta}{\theta}\right)^{2/3}\left[5 + \sin^{-2}\left(\frac{\theta}{2}\right)\left(\frac{\sin\theta}{\theta} - 1\right)\right] \tag{4-36}$$

where θ is the central angle in radians, as shown in Figure 3-1 (Yen, 2004).

4.3.4 Muskingum-Cunge Method

The Muskingum-Cunge method is a variation of the diffusion wave model that is based on a finite difference extension of the well known Muskingum method. While there are more sophisticated and accurate methods available to predict wave propagation, the Muskingum-Cunge method is appealing and widely used due to is simplistic nature and modest computational demand. Considering a similar time and space grid to that of Figure 4-2, discharge can be expressed as

$$Q_{i+1}^{k+1} = C_1 Q_i^{k+1} + C_2 Q_i^k + C_3 Q_{i+1}^k \tag{4-37}$$

where the coefficients C_1, C_2, and C_3, referred to as Muskingum coefficients, sum to unity and are defined as

$$C_1 = \frac{\Delta t - 2KX}{2K(1 - X) + \Delta t} \tag{4-38}$$

$$C_2 = \frac{\Delta t + 2KX}{2K(1-X) + \Delta t} \tag{4-39}$$

$$C_3 = \frac{2K(1-X) - \Delta t}{2K(1-X) + \Delta t} \tag{4-40}$$

In the original Muskingum method, Cunge (1969) showed that if K and X are constant over time for individual channels, Equation 4-37 is an approximation of the kinematic wave model. In this case, the constants are typically calibrated using flow measurements. Cunge (1969) further showed that Equation 4-37 is an approximation of a modified diffusion wave model if K and X are evaluated based on channel properties and flow characteristics as

$$K = \frac{\Delta x}{c_k} \tag{4-41}$$

which represents the time propagation of a given discharge along channel length Δx, and

$$X = \frac{1}{2}\left(1 - \frac{Q}{Bc_k S_o \Delta x}\right) \tag{4-42}$$

where Q refers to a reference discharge. Thus, after computing the Muskingum coefficients, Equation 4-37 can be solved for each point on the time and space grid. It is important to note that while the apparent difference between the original Muskingum method and the Muskingum-Cunge method narrows to the manner in which K and X is computed, the two methods are fundamentally different. The former is a hydrologic method that critically relies upon a relationship to channel storage. The latter is referred to as a hydraulic method that is based on an approximate solution of the St. Venant equations.

For numerical stability, the value of X should range from 0.0 to 0.5. In addition, the time step and distance step should be selected so that (Ponce and Theurer, 1982)

$$\Delta t \le \frac{T_r}{M} \tag{4-43}$$

where $M \ge 5$, and

$$\Delta x \leq \frac{1}{2}\left(c_k \Delta t + \frac{Q}{B c_k S_o} \right) \tag{4-44}$$

which tends to be slightly conservative. Although no limitation is necessarily imposed via the Courant number, a value of C_n close to unity is suggested for reasons of accuracy.

If the reference discharge, and thus kinematic wave celerity since $c_k = f(Q)$, is held constant in determination of K and X, the resulting model is similar to the original Muskingum method with the same parameters and is referred to as the constant-parameter method (Dooge, 1973). Alternatively, a nonlinear, variable-parameter method is created by permitting these terms to vary with newly-computed discharges. At least two options are available with respect to the nonlinear case. First, the reference discharge and c_k can be updated for each time ordinate. Updated values could be based on an average or other fraction, a, of the inflow rate for a particular time interval, where $0 \leq a \leq 1.0$ (Premual, 1992). For example, Samani and Jebelifard (2003) used a nonlinear regression technique to show that for a circular conduit of diameter D,

$$a = \frac{0.0163 D^{0.5147}}{n^{0.5724} S_o^{0.0158} Q_i^{0.2285}} \tag{4-45}$$

where Q_i is the inflow at the upstream end of the channel. It is important to note, however, that the exponents and coefficient used in Equation 4-45 are specific to particular simulation data and may not be universally applicable. For potential further improvement, the second option involves allowing parameters to vary with cross section location, as well as time. The reference discharge and wave celerity can be based on a three-point average of known conditions (i.e., Q_i^{k+1}, Q_i^k, and Q_{i+1}^k) or an iterative, four-point average that considers unknown conditions (i.e., Q_i^{k+1}, Q_i^k, Q_{i+1}^k, and Q_{i+1}^{k+1}) (Ponce and Yevjevich, 1978; Ponce and Chaganti, 1994). For the latter, the three-point average can be used to obtain an initial estimate in the iterative computation. While tests show that four-point schemes are slightly better (Tang et al., 1999), McCuen (2005) notes that the three-point and four-point methods should not typically yield significantly different computed hydrographs; thus the iterative technique is not usually worth the additional computational effort.

The Muskingum-Cunge method has two significant advantages over the kinematic wave model. First, the solution is obtained through a linear algebraic equation rather than finite difference or characteristic approximations of partial differential equations. This allows derivation of hydrographs at required cross sections instead of requiring a solution over the entire length of channel for each time increment. Second, solution of Equation 4-37 should show less wave attenuation than the kinematic wave method, thus permitting a more flexible choice of time and space intervals (Chow et al., 1988). The disadvantages of the method are that it cannot directly accommodate downstream disturbances that propagate upstream (i.e., backwater effects) and that it may not accurately simulate the hydrograph at a downstream boundary when large variations in kinematic wave celerity occur. The latter implies that time and space increments (i.e., Δx and Δt) should be small at the system outfall. Finally, the method should generally be used for flows peaking in more than approximately two hours since fast-rising hydrographs may not be adequately simulated. In this regard, Ponce (1989) suggests the following as a criterion for applicability:

$$T_r S_o \sqrt{\frac{g}{y}} \geq 15 \qquad (4\text{-}46)$$

Due to its inherent simplicity and computational efficiency, Muskingum-Cunge routing algorithm is well suited and widely used for analyzing unsteady flow conditions in sewer systems. A recent study of the numerical modeling aspects of unsteady flow in simple and compound channels found that such an approach, using the same laboratory test data, compares favorably in terms of overall performance to more complex unsteady flow dynamic routing models that solve the complete form of the St. Venant equations in one dimension (Chatila, 2003).

4.4 SEWER SURCHARGE

A surcharge condition exists when a sewer flows full under pressure, potentially as a result of an extreme rain event or an under-designed conduit. Unsteady surcharge flows can be simulated using standard transient pipe flow techniques or the Preissmann hypothetical piezometric open-slot approach. For the transient pipe flow approach, the effective flow area remains unchanged (i.e., $\partial A/\partial t = 0$). Furthermore, if the sewer pipe maintains a constant cross-section and flows full over its length, L, then $\partial V/\partial x = 0$. Thus, the St. Venant equations reduce to

$$Q = VA \qquad (4\text{-}47)$$

$$H_u - H_d - h_l - S_f L - \frac{L}{gA}\frac{\partial Q}{\partial t} = 0 \qquad (4\text{-}48)$$

where H_u and H_d are total energy head at the entrance and exit, respectively, and h_l represents total form losses (see Chapter 3).

The piezometric open-slot approach hypothetically assumes that a narrow, continuous slot, as shown in Figure 4-3, is attached to the pipe crown. This concept allows the transformation of a pressurized pipe flow problem into an open channel problem by introducing a virtual free surface (Yen, 2004). Some of the benefits of this approach are that it is not necessary to periodically switch between pipe surcharge and open channel flow equations and that it can easily account for cases in which only a portion of a sewer length is surcharged. However, for sewer networks comprised of many reaches or spatial elements, computational demand may be high.

The critical aspect of the hypothetical open-slot approach is establishing the width of the slot. The width should not be so large as to introduce error in volumetric computations, but it should also not be too narrow in order to avoid numerical difficulties associated with rapid pressure surges. A theoretical basis for determining slot width is to establish a size such that the wave celerity in a slotted sewer of diameter D is the same as the surge celerity, c, of water in an elastic pipe without a slot. In this case, the width, b, is expressed as

$$b = \frac{\pi g D^2}{4c^2} \tag{4-49}$$

Note that in spite of the value obtained from Equation 4-49, Cunge et al. (1980) recommend using a minimum slot width of 0.4 in (1 cm).

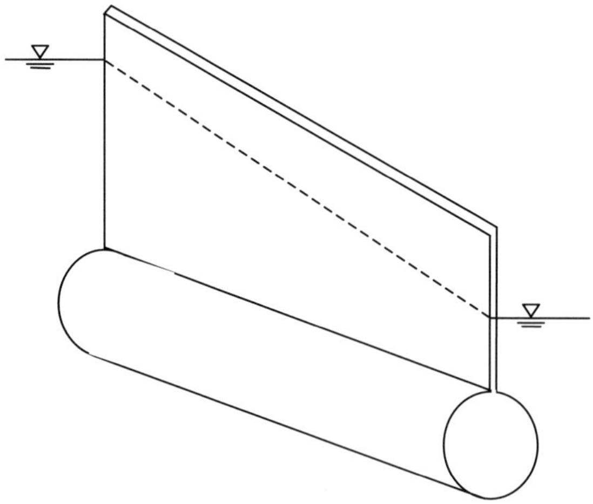

Figure 4-3: Preissmann hypothetical piezometric open slot

4.5 SOLVED PROBLEMS

Problem 4.1 Method of characteristics

In a wide rectangular channel with bed slope of 0.0003 and Manning's n of 0.022, the following simultaneous observations were made at points A and B during a passage of a flood wave. Point B is located 6 km downstream of point A. Use the method of characteristics to determine the location at which the characteristic curves intersect, and the flow velocity and depth at the point of intersection.

Point	Velocity (m/s)	Depth (m)
A	0.85	1.2
B	0.75	1.1

Solution

Considering Figure P4-1, point P represents the intersection point of the forward characteristics (*C^+*) and the backward characteristics (*C^-*) curves.

To determine the location of point P, one has to solve Equations 4-10 and 4-11 using slopes at points A and B for the C^+ and C^- curves, respectively, as

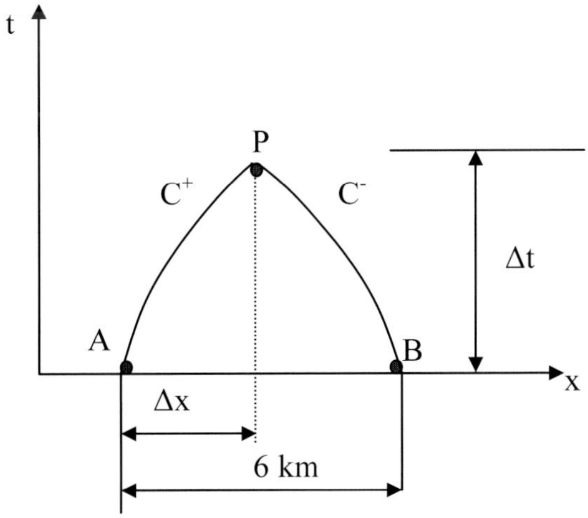

Figure P4-1

$$\left(\frac{dx}{dt}\right)_A = V_A + c_A, \text{ and } \left(\frac{dx}{dt}\right)_B = V_B - c_B$$

Using a difference approximation at point A,

$$\Delta t = \frac{\Delta x}{V_A + c_A}$$

and at point B,

$$\Delta t = \frac{(\Delta x - 6000)}{V_B - c_B}$$

Table P4-1 summarizes the calculations required to obtain Δx and Δt.

Table P4-1

Point	V	y	c	dx/dt	S_f	V±2c
A	0.85	1.2	3.759	4.609	0.00051	8.368
B	0.75	1.1	3.445	-2.695	0.00056	-6.14

where V and y are velocity and flow depth at the two points; c is wave celerity calculated as $\sqrt{gy}$; dx/dt is slope of the respective curves at points A and B, and is computed using Equations 4-10 and 4-11, respectively; S_f is slope of the energy grade line and is computed from Manning's equation as

$$S_f = \frac{n^2 V^2}{y^{4/3}}$$

Therefore, at point A,

$$\Delta t = \frac{\Delta x}{4.609},$$

And at point B,

$$\Delta t = \frac{-(\Delta x - 6000)}{2.695}$$

Solving for Δt and Δx

$$\Delta t = 304.81\, sec \text{ and } \Delta x = \underline{\underline{1{,}404.88\, m}}$$

Along the curves, the compatibility equations (i.e., Equations 4-8 and 4-9) must be satisfied. Expressing the equations in difference form, along the C^+ curve from A to P

$$(V+2c)_P-(V+2c)_A=\frac{\Delta t\times g}{2}\left(2S_0-S_{fA}-S_{fp}\right) \qquad \text{(P4:1-1)}$$

and along the C^- curve from B to P

$$(V-2c)_P-(V-2c)_B=\frac{\Delta t\times g}{2}\left(2S_0-S_{fB}-S_{fA}\right) \qquad \text{(P4:1-2)}$$

Subtracting Equation P4:1-2 from Equation P4:1-1

$$4c_P=(V+2c)_A-(V-2c)_B+\frac{\Delta t\times g}{2}\left(S_{fB}-S_{fA}\right)$$

$$=8.368-(-6.14)+\frac{304.81\times 9.81}{2}(0.00056-0.00051)$$

$$=14.583$$

Therefore, $c_p = 3.646$ m/s. Depth of flow at point P, y_P, is calculated as

$$y_P=\frac{c_P^2}{g}=\frac{3.646^2}{9.81}=\underline{\underline{1.355\,m}}$$

Note that the above approach is essentially the same as applying Equations 4-12 through 4-15.

The characteristic equation for either of the curves (i.e., C^+ or C^-) may be used to evaluate the velocity at point P. Using Equation P4:1-1, and substituting corresponding values of the variables

$$V_P+7.292-8.368=1{,}495.1\times\left(.0006-0.00051-S_{fp}\right)$$

Knowing that $S_{fP}=0.00043V_P^2$, one can obtain

$$0.6433V_P^2+V_P-1.21=0$$

Therefore, $V_P=\underline{\underline{0.799\,m/s}}$.

Note that location of point P was originally determined by linear approximation using the slopes at points A and B. If slope (i.e. *dx/dt*) at point P is computed using the information obtained at the point (i.e., V_p and c_P), as shown below, the two slopes are different.

$$\left(\frac{dx}{dt}\right)_P = V_P \pm c_P = \text{4.445 along } C\text{+ curve, and –2.847 along the } C^{-} \text{ curve.}$$

When compared to the 4.609 (i.e., $V_A + c_A$) and -2.695 (i.e., $V_B - c_B$) used initially, the slopes computed at P are smaller.

Comments: Using information at A and B only to determine the location and flow properties at P can lead to considerable error particularly for conditions where the flow depth is changing relatively fast due to flood waves. Under those scenarios, to improve the velocity gradient (*dx/dt*) approximation, the average gradient between A and P for the forward characteristic and B and P for the backward characteristic should be used. The computational accuracy should be improved by carrying out successive iterations until the previous and the present slopes match.

Problem 4.2 Kinematic wave celerity

A triangular channel has a slope of 0.0008 ft/ft, a Manning roughness coefficient of 0.032, and side slopes of 3H:1V. Determine the kinematic wave parameters, α_k and m, along with an associated expression for discharge that is similar in form to Equation 4-26.

Solution

From geometry

$$A = 3y^2$$

$$P = 2y\sqrt{1+3^2} = 6.32y$$

and

$$R = \frac{A}{P} = 0.47\,y$$

Substituting into the Manning Equation, assuming U.S. customary units, yields

$$Q = \frac{1.49}{0.032}\left(3y^2\right)\left(0.47y\right)^{2/3}\left(0.0008\right)^{1/2} = 2.39y^{8/3}$$

Then, knowing that $y = \sqrt{A/3}$ from above,

$$Q = 0.55A^{4/3}$$

which has the form of Equation 4-26. Thus, α_k is 0.55 and m is 4/3.

Problem 4.3 Kinematic wave celerity

Consider an 8-ft diameter circular sewer having a Manning roughness coefficient of 0.015 and laid on a slope of 0.001 ft/ft. At a discharge of 140 cfs, the uniform flow depth is 4.3 ft. For these conditions, compute the speed associated with a kinematic wave.

Solution

The central angle θ is computed by

$$\theta = 2\cos^{-1}\left(1 - \frac{2y}{D}\right) = 2\cos^{-1}\left(1 - \frac{2(4.3)}{8}\right) = 3.29\ rad$$

Then by Equation 4-36,

$$c_k = \frac{0.132(1.49)}{0.015}(0.001)^{1/2}(8)^{2/3}\left(1 - \frac{\sin 3.29}{3.29}\right)^{2/3} \times\left[5 + \sin^{-2}\left(\frac{3.29}{2}\right)\left(\frac{\sin 3.29}{3.29} - 1\right)\right] = 10.3\ fps$$

Comments: Strictly speaking, this wave speed applies to only the depth and discharge given in the problem statement. As discharge changes spatially and temporally, the celerity should be updated.

Problem 4.4 Muskingum-Cunge method

The channel described in Problem 4.1 is subjected to the inflow hydrograph below. Using a time and space increment (i.e., Δt and Δx) of 0.25 hours and 2,000 ft, respectively, use the Muskingum-Cunge method to calculate the discharge at the end of the 2,000-ft long channel. Use the constant-parameter method with a reference discharge of 60 cfs (i.e., peak discharge). How does the result change if a mid-range value of 35 cfs is used as the reference discharge?

Time (hr)	0.0	0.25	0.5	0.75	1.0	1.25	1.5	1.75	2.0
Inflow (cfs)	10	16	31	50	58	60	54	42	32
Time (hr)	2.25	2.5	2.75	3.0	3.25	3.5	3.75	4.0	4.25
Inflow (cfs)	25	20	17	15	13	12	11	10	10

Solution

Rearranging the result from Problem 4.1,

$$A = \left(\frac{Q}{0.55} \right)^{3/4}$$

Then, at a reference discharge of 60 cfs, A is 33.76 ft^2. Also,

$$y = \sqrt{\frac{A}{3}} = \sqrt{\frac{33.76}{3}} = 3.35\ ft$$

and top width is defined by

$$B = 2zy = 2(3)(3.35) = 20.10\ ft$$

The flow velocity can be computed as

$$V = \frac{Q}{A} = \frac{60}{33.76} = 1.78\ fps$$

so that c_k becomes

$$c_k = mV = \frac{4}{3}(1.78) = 2.37\ fps$$

From Equations 4-41 and 4-42,

$$K = \frac{2000}{2.37} = 843.9\ sec = 0.23\ hrs$$

and

$$X = \frac{1}{2}\left(1 - \frac{60}{(20.10)(2.37)(0.0008)(2000)} \right) = 0.106$$

Substitution of K and X into Equations 4-38 through 4-40 yields $C_1 = 0.304$, $C_2 = 0.452$, and $C_3 = 0.244$. A summary of the solution that uses these values within Equation 4-37 is shown in Table P4-4. The Table provides the outflow hydrograph in Column 6 for a reference discharge of 60 cfs. If a reference discharge of 35 cfs is used instead of 60 cfs, the values of C_1, C_2, and C_3 are 0.214, 0.513, and 0.273, respectively. The new outflow hydrograph is shown in Column 7. Figure P4-4 shows a comparison of the two alternative outflow hydrographs. An important aspect to reflect upon, however, is the assumption of a constant reference discharge and, thus, wave celerity. This simplification may cause a failure to fully capture the physical flow behavior, particularly the steepening and flattening on the rising and falling limbs of the hydrograph. Although the general tabular approach would remain the same, an improved approach would incorporate updated values of kinematic wave celerity and Muskingum coefficients as a function of computed discharge.

Table P4-4

(1)	(2)	(3)	(4)	(5)	(6)	(7)
Time (hr)	Inflow (cfs)	$C_1 \times Q_i^{k+1}$	$C_2 \times Q_i^k$	$C_3 \times Q_{i+1}^k$	Outflow (cfs) ($Q = 60$ cfs)	Outflow (cfs) ($Q = 35$ cfs)
0.00	10	-	-	-	10.0	10.0
0.25	16	4.86	4.53	2.44	11.8	11.3
0.50	31	9.42	7.25	2.89	19.6	17.9
0.75	50	15.20	14.04	4.77	34.0	31.5
1.00	58	17.63	22.65	8.30	48.6	46.7
1.25	60	18.24	26.27	11.85	56.4	55.3
1.50	54	16.42	27.18	13.75	57.4	57.4
1.75	42	12.77	24.46	13.99	51.2	52.4
2.00	32	9.73	19.03	12.50	41.3	42.7
2.25	25	7.60	14.50	10.07	32.2	33.4
2.50	20	6.08	11.33	7.85	25.3	26.2
2.75	17	5.17	9.06	6.16	20.4	21.1
3.00	15	4.56	7.70	4.98	17.2	17.7
3.25	13	3.95	6.80	4.21	15.0	15.3
3.50	12	3.65	5.89	3.65	13.2	13.4
3.75	11	3.34	5.44	3.22	12.0	12.2
4.00	10	3.04	4.98	2.93	11.0	11.1
4.25	10	3.04	4.53	2.67	10.2	10.3

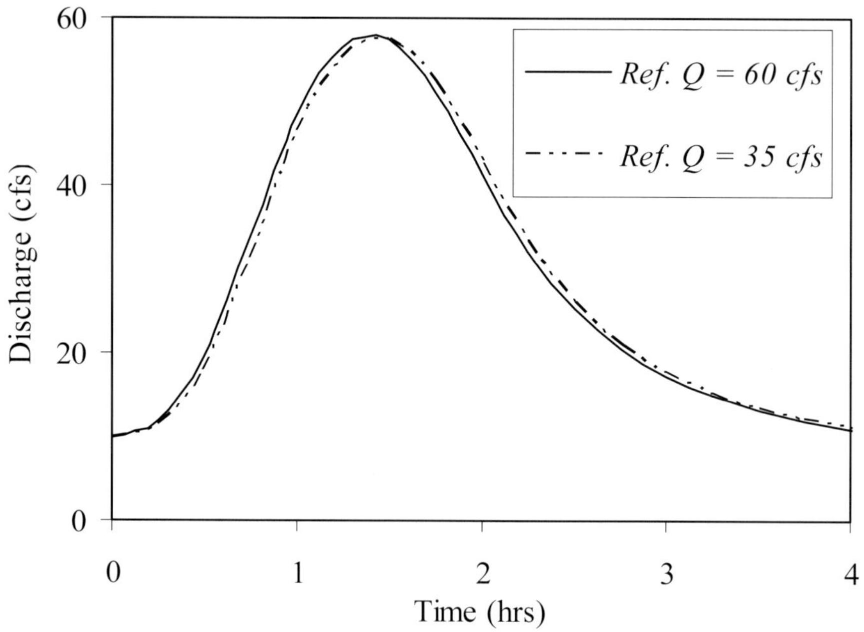

Figure P4-4

Problem 4.5 Muskingum-Cunge method

Hypothetically assume that a 3-mile length of the sewer described in Problem 4.3 exists. The inflow hydrograph at the upstream end has a peak discharge of 250 cfs at a time to peak of 2 hours and has a base time of 5 hours. Assuming a triangular hydrograph with a base (i.e., minimum) flow of 30 cfs, determine the outflow peak discharge and time of occurrence using the Muskingum-Cunge method.

Solution

For simplicity and demonstration purposes, the constant-parameter method is used with a reference discharge of 140 cfs, which is midway between the base and peak discharges. At 140 cfs, c_k assumes a value of 10.3 fps (see Problem 4.3). Based on Equation 4-43, a value of Δt equal to 0.25 hrs is tentatively chosen, and at the reference discharge (and depth of 4.3 ft), top width is computed as

$$B = D\sin\frac{\theta}{2} = 8\sin\frac{3.29}{2} = 7.8\ ft$$

Then, from Equation 4-44,

$$\Delta x \leq \frac{1}{2}\left((10.3)(0.25 \times 3600) + \frac{140}{(7.8)(10.3)(0.001)} \right) = 5{,}506\ ft$$

Therefore, three 1-mile (i.e., 5,280 ft) long computational reaches can be used. The Courant number for these time and space increments is

$$C_n = \frac{c_k \Delta t}{\Delta x} = \frac{(10.3)(0.25 \times 3600)}{5280} = 1.75$$

which is greater than one. Further adjustments of Δt and Δx could be undertaken to improve accuracy since the criterion given by Equation 4-44 is somewhat conservative. For illustration purposes, however, these values will be used. From Equations 4-41 and 4-42,

$$K = \frac{5280}{10.3} = 512.6\ sec = 0.14\ hrs$$

and

$$X = \frac{1}{2}\left(1 - \frac{140}{(7.8)(10.3)(0.001)(5280)} \right) = 0.34$$

Substituting K and X into Equations 4-38 through 4-40 thus yields $C_1 = 0.356$, $C_2 = 0.794$, and $C_3 = -0.150$, which sum to unity as required. Table P4-4 represents the stepwise solution of Equation 4-37. Column 2 represents the upstream hydrograph, interpolated from the problem statement. Then, for example, at $k = 0.25$ hours and $i = 0$, corresponding to the upstream end,

$$Q_{i+1}^{k+1} = (0.356)(85.0) + (0.794)(57.5) + (-0.15)(39.8) = 70\ cfs$$

Note that the outflow hydrograph from an upstream reach becomes the inflow for the next downstream reach. Within Table P4-5, only final results are shown for downstream reaches (x = 2 and 3 mi), without intermediate computations. Based on the results shown, the peak flow at a distance of 3 miles downstream is 241.8 cfs and occurs at 2.5 hours, which is of course dependent on the time interval of 0.25 hours. As with Problem 4-3, improvements may be made by assuming a temporally and spatially varying kinematic wave celerity that is based on computed flow rates.

Table P4-5

(1)	(2)	(3)	(4)	(5)	(6)	(7)	(8)
Time (hr)	Inflow (cfs)	$C_1 \times Q_i^{k+1}$	$C_2 \times Q_i^k$	$C_3 \times Q_{i+1}^k$	Outflow (cfs) (x=1 mi)	Outflow (cfs) (x=2 mi)	Outflow (cfs) (x=3 mi)
0.00	30.0	-	-	-	30.0	30.0	30.0
0.25	57.5	20.47	23.82	-4.50	39.8	33.5	31.2
0.50	85.0	30.26	45.66	-5.97	70.0	51.5	40.2
0.75	112.5	40.05	67.49	-10.49	97.1	82.4	64.2
1.00	140.0	49.84	89.33	-14.56	124.6	109.1	94.6
1.25	167.5	59.63	111.16	-18.69	152.1	136.7	121.1
1.50	195.0	69.42	133.00	-22.81	179.6	164.2	148.9
1.75	222.5	79.21	154.83	-26.94	207.1	191.7	176.3
2.00	250.0	89.00	176.67	-31.06	234.6	219.2	203.8
2.25	231.7	82.47	198.50	-35.19	245.8	240.9	229.2
2.50	213.3	75.95	183.94	-36.87	223.0	238.4	241.8
2.75	195.0	69.42	169.39	-33.45	205.4	214.4	229.4
3.00	176.7	62.89	154.83	-30.80	186.9	197.4	206.1
3.25	158.3	56.37	140.27	-28.04	168.6	178.8	189.5
3.50	140.0	49.84	125.72	-25.29	150.3	160.5	170.7
3.75	121.7	43.31	111.16	-22.54	131.9	142.2	152.5
4.00	103.3	36.79	96.60	-19.79	113.6	123.9	134.1
4.25	85.0	30.26	82.05	-17.04	95.3	105.5	115.8
4.50	66.7	23.73	67.49	-14.29	76.9	87.2	97.5
4.75	48.3	17.21	52.93	-11.54	58.6	68.9	79.1
5.00	30.0	10.68	38.38	-8.79	40.3	50.5	60.8

REFERENCES CITED

Chaudhry, M.H., *Open Channel Flow*, Prentice-Hall, London, 1993.

Chatila, J.G., " Muskingum Method, EXTRAN and ONE-D for Routing Unsteady Flows in Open Channels," *Canadian Water Resources J.*, vol. 28, no.3, 481-498, 2003.

Chow, V.T., D.R. Maidment, and L.W. Mays, *Applied Hydrology*, McGraw-Hill, New York, NY, 2001.

Cunge, J.A., "On the Subject of Flood Propagation Method (Muskingum Method)," *J. of Hydraulic Research*, IAHR, vol. 7, no. 2, 205-230, 1969.

Cunge, J.A., F.M. Holly, and A. Verwey, *Practical Aspects of Computational River Hydraulics*, Pitman, London, 1980.

Dooge, J.C.I., "Linear Theory of Hydrologic Systems," *USDA Tech. Bull. 1468*, U.S. Department of Agriculture, Washington, DC, 1973.

McCuen, R.H., *Hydrologic Design and Analysis*, 3rd ed., Upper Saddle River, NJ, 2005.

Ponce, V.M., *Engineering Hydrology*, Prentice Hall, Englewood Cliffs, NJ, 1989.

Ponce, V.M., and P.V. Chaganti, "Variable-Parameter Muskingum-Cunge Method Revisited," *J. of Hydrology*, vol. 162, 433-439, 1994.

Ponce, V.M., and F.D. Theurer, "Accuracy Criteria in Diffusion Routing," *J. of the Hydraulic Div.*, ASCE, vol. 108, no. HY6, 747-757, 1982.

Ponce, V.M., and V. Yevjevich, "Muskingum-Cunge Method with Variable Parameters," *J. of Hydraulic Div.*, ASCE, vol. 104, no. HY12, 1663-1667, 1978.

Perumal, M., "Multilinear Muskingum Flow Routing Method," *J. Hydrology*, vol. 133, 259-272, 1992.

Samani, H.M.V., and S. Jebelifard, "Design of Circular Urban Storm Sewer Systems Using Multilinear Muskingum Flow Routing Method," *J. of Hydraulic Engrg.*, ASCE, vol. 129, no. 11, 832-838, 2003.

Sturm, T.W., *Open Channel Hydraulics*, McGraw-Hill, New York, NY, 2001.

Tang X-N., D.W. Knight, and P.G. Samuels, "Volume Conservation in Variable Parameter Muskingum-Cunge Method," *J. of Hydraulic Engrg.*, ASCE, vol. 125, no. 6, 610-620, 1999.

Yen, B.C., "Hydraulics of Sewer Systems," in *Urban Stormwater Management Tools*, ed. by L.W. Mays, McGraw Hill, New York, NY, 2004.

CHAPTER FIVE
Water Quality Analysis

5.1 INTRODUCTION

Once the hydraulic investigation of a sewer system has been completed, a water quality analysis can be undertaken to determine the distribution of constituent concentrations and to identify areas of potential concern. Water quality broadly refers to the amount of suspended (e.g., sediment and finer particulate matter) and dissolved solids, dissolved gases, and microorganisms being transported by flow.

Human influence on the quality of storm and sanitary sewer flows is significant. Consider, for example, the pollutant loads associated with human or industrial wastes, erosion resulting from agricultural and construction practices, the application of fertilizers and pesticides, litter, and wash-off from streets and parking lots. It is an important issue, particularly with respect to receiving waters, since quality affects the availability of water for various uses and, in a broader scope, a perceived quality of life.

5.2 FLOW AGE

Age of flow represents the time from when flow enters a sewer system or particular conduit to when it exits. Since constituent concentrations depend on detention time in a system, age can be used as a general indication of overall quality. Using zero order kinetics, age can be expressed as

$$T_j = T_{j-1} + \frac{L}{V} \tag{5-1}$$

where T_j represents the age of water at the *j-th* cross-section or junction; L is the conduit length from $j-1$ to j; and V is the velocity of flow in the corresponding conduit. This parameter is useful to address important water quality and safety issues such as generation of hydrogen sulfide that may occur in sanitary sewer systems, manifesting itself in corrosion and odor issues.

5.3 SOURCE TRACE

Source tracing refers to determining the percentage of flow originating from a particular location in the sewer. Considering N_s sources, tracing requires $N_s - 1$ water quality simulations. Individual evaluations are made with a fictitious conservative substance supplied from a given location (i.e., n_s) in the network and having a concentration of 100 units (Boulos et al., 2004). The load is permitted to move through the network, and the concentration at any downstream location is the percentage contribution from source n_s. It should be evident that tracing can be especially useful in evaluating the spatial and temporal extent of influence of a particular source and the degree of mixing that occurs through a system.

Source tracing is very useful in sewer systems modeling and applications, and could be used for (1) tracking changes in sewage flow contribution and associated constituents over space and time; (2) predicting impact of industrial and commercial waste discharges on performance of wastewater treatment plants; (3) determining contaminant level that causes a wastewater treatment plant to be in violation of its discharge permits; (4) and developing appropriate user charges based on wasteloads and level of contaminant.

5.4 TRANSPORT OF CONSERVATIVE CONSTITUENTS

Constituents will travel down the length of a sewer pipe primarily through the process of advection, whereby a substance moves in response to externally applied pressure forces or as a result of gravity. The implication is that the movement will occur in the same direction as, and with a velocity equal to, the bulk fluid. To a smaller extent, transport is also affected by diffusion and dispersion. Diffusion is transport caused by molecular movement (i.e., Brownian motion), while dispersion refers to transport caused by local mixing and variable velocity regimes. For turbulent flow through conduits, as is generally the case with sewer flows, transport is dominated by advection.

Advective transport of conservative (inert) constituents, those not subject to reactions or biochemical degradation, can be described by (Boulos, et al., 2004)

$$\frac{\partial C}{\partial t} + V\frac{\partial C}{\partial x} = 0 \tag{5-2}$$

where V is the bulk flow velocity, determined via hydraulic analysis; C is constituent concentration; t is time; and x longitudinal distance along the sewer. Equation 5-2 can be written once for each length of conduit in a system.

At manholes and junctions, assuming complete and instantaneous mixing of loads, a mass balance yields

$$C_e = \frac{\sum_{i=1}^{I} C_i Q_i}{\sum_{i=1}^{I} Q_i} \tag{5-3}$$

where C_e represents the constituent concentration at the junction exit, typically expressed in mg/L; Q is volumetric flow rate; and I refers to the total number of conduits entering the junction.

Conservative constituent transport can also be simulated using hydrologic or hydraulic routing methods such as the Muskingum or Muskingum-Cunge methods (see Chapter 4) (McCuen, 2005). In this case, a loadograph that describes the temporal distribution of a pollutant load is routed through a sewer. Pollutant load, L, is directly related to concentration through volumetric flow rate, or

$$L = K_p CQ \tag{5-4}$$

where K_p is a constant equal to 5.39 if L, C, and Q are expressed in lb/day, cfs, and mg/L, respectively. In S.I. units, K_p assumes a value of 1.0 when L, C, and Q are expressed in g/sec, m^3/s, and mg/L, respectively. It is important to note that while Equation 5-4 is written in terms of single-valued parameters, they actually vary spatially and temporally. Once a loadograph has been evaluated at an upstream sewer section, it can be used with a routed hydrograph to determine the distribution of pollutant at a downstream section.

The ability to model pollutant transport in sewer collection systems is useful in determining the amount of pollutant that is transported to the wastewater treatment plant and assessing its impacts on receiving waters.

5.5 SEDIMENT MOVEMENT

Sewer sediments can be defined as any settleable particulate matter found in storm or sanitary systems (Butler et al., 2003). For sanitary sewers, domestic wastewater typically contains between 5 and 20 mL/L of settleable material (Metcalf and Eddy, 1991). For stormwater systems, a significant amount of sediment enters through erosion processes and wash-off of surface deposits from streets and other paved areas. For pervious areas, the quantity of solids can be computed using the universal soil loss equation (USLE) (Wischeimer and Smith, 1958; Huber and Dickinson, 1988), or one of its proposed variants (e.g., the MUSLE or RUSLE). The USLE is expressed as

$$E = 2000(AR_r KL_S CP_E) \tag{5-5}$$

where E is the weight of solids in pounds contained in stormwater runoff over a specified period Δt; A is basin area in acres; and R_r is a rainfall factor that depends on geographic location and accounts for erosive force of the rainfall, which can be evaluated by

$$R_r = i_{30}(9.16 + 3.31 \log i) i \Delta t \tag{5-6}$$

where i is rainfall intensity in in/hr during Δt; i_{30} is the maximum average 30-minute rainfall intensity for the storm; and Δt is in hours. K is a soil erodibility factor in tons/acre and ranges from 0.10 for sands to more than 0.35 for silt loam or silty clay loam. L_S is a dimensionless length-slope factor that reflects the effects of basin length and steepness and is evaluated empirically as

$$L_S = \sqrt{L}\left(0.0076 + 0.53 S_o + 0.76 S_o^2\right) \tag{5-7}$$

where L is the overland flow length in feet and S_o is the average surface slope for the runoff. C is a dimensionless land cover and management factor that assumes a value of 1.0 for bare soil. For areas covered more than 95 percent by short vegetation, a reasonable value is 0.003, and for forests having significant canopy, C may be as small as 0.0001. P_E is a dimensionless erosion control practice factor that accounts for tillage and conservation practices. For urbanized areas, however, P_E can be taken as 1.0. It should be noted that the USLE was originally developed to estimate long term (e.g., annual) erosion rates. Therefore, application to shorter durations and single storm events provides only approximate results.

For impervious, urban areas, experimental work by Sartor and Boyd (1972) indicates that the weight of solids washed off at time t is

$$E = E_o e^{-kr} \tag{5-8}$$

where E_o is the initial weight of solids, and r is the total rainfall depth in inches, obtained by multiplying the runoff rate by its duration. The proportionality constant k is often taken as 4.6 in^{-1} and can range from 0.052 in^{-1} to 6.6 in^{-1} (Sonnen, 1980). The amount of wash off that occurs during the storm duration is then the difference between E_o and E.

Once sediment has entered the sewer, its movement can occur in three phases; solids are entrained by the fluid when shear stress reaches a critical value, then are transported either in suspension (i.e., lighter material) or by rolling, sliding, or jumping along the conduit invert as bed load (i.e., heavier particles), and then are potentially deposited downstream. Deposition occurs when the quantity of available sediment exceeds the transport capacity of the fluid. If left uncontrolled, excess deposition can lead to clogging, surcharge, and added overflows. In sanitary sewers, deposition poses additional problems such as the promotion of anaerobic conditions and generation of hydrogen sulfide. Furthermore, since a number of constituents (e.g., organics, metals, and hydrocarbons) are capable of becoming sorbed to sediments, reentrainment and transport of already deposited material (e.g., during dry weather) can cause increased pollutant loads within sewers.

Before discussing quantitative approaches for simulating sediment movement in sewers, it is important to contextually understand the applicability of existing relationships and our current state of knowledge. Generally speaking, traditional relationships for fluvial sediment transport are quite limited in the analysis of sewers. De Sutter et al. (2003) and Ackers et al. (1996) experimentally documented the poor performance of several existing transport equations when applied to in-sewer sediment transport. The limitations are, in part, a result of the rigid, circular cross-sectional shape used for most sewers and the nature of associated flows, especially when compared to the characteristics of rivers and streams, and a result of the possible existence of cohesive matter, particularly in the case of combined and sanitary systems. Because of the complexities involved, methods for modeling sewer sediments are incomplete and will require further research. As a result, a Task Committee was appointed in 1999 by the American Society of Civil Engineers to advance knowledge of sediment movement in sewers (Delleur, 2003). It is hoped that through the work of this Task Committee and other research being conducted around the world, existing methods will evolve and become more accurate for both sediment prediction and the evaluation of minimum (i.e., self cleansing) and maximum flow velocities.

The most recent and comprehensive work for sediment movement in circular sewers includes that of the British Construction Industry Research and Information Association (CIRIA). A 1996 CIRIA report (Ackers et al., 1996) recommends that the Macke (1982) formula be used to simulate suspended sediment at the limit of deposition (i.e., no deposition). The formula can be expressed as (Delleur, 2001)

$$C_v = \frac{f_c^3 V_L^3}{30.4(S-1)w^{1.5} A} \tag{5-9}$$

where C_v is the volumetric sediment concentration (i.e., ratio of volumetric discharge of sediment to volumetric discharge of fluid); f_c is the composite Darcy-Weisbach friction factor for sediment-laden flow through sewer pipes, evaluated as

$$f_c = \frac{P_w f_w + P_b f_b}{P_w + P_b} \tag{5-10}$$

where P refers to wetted perimeter, and subscripts w and b refer to the pipe wall and sediment bed. The term f_w corresponds to a clean-pipe friction factor, which is determined using the Colebrook-White equation (see Chapter 3) or similar method. V_L in Equation 5-9 is the limiting flow velocity without deposition (i.e., required to avoid deposition); S is the specific gravity of sediment particles; A is cross-sectional area of flow; and w refers to the particle settling velocity. The expression is written in S.I. units and is valid when shear stress, τ, is greater than 1.07 N/m^2. The average shear stress can be approximated by

$$\tau = \gamma R S_f \tag{5-11}$$

where γ is specific weight of the fluid-sediment mixture; R is the hydraulic radius; and S_f refers to the friction slope, or slope of the energy grade line. Alternatively, if channel roughness is taken into consideration,

$$\tau = \frac{f_b}{8} \rho V^2 \tag{5-12}$$

where ρ refers to fluid density.

For sewer systems in which some deposition is expected, the CIRIA report suggests use of the modified Ackers-White equation (Ackers, 1991; Ackers et al., 1996), expressed as (Delleur, 2001)

$$C_v = J\left[\frac{W_e R}{A}\right]^{\alpha}\left[\frac{d}{R}\right]^{\beta} f_c^{\lambda}\left[\frac{V}{\sqrt{g(S-1)R}} - K f_c^{\delta}\left(\frac{d}{R}\right)^{\varepsilon}\right]^{m} \tag{5-13}$$

The effective width of the sediment bed, W_e can be expressed as a function of the sediment bed width as

$$W_e = \left[0.2 + 3.33\left(\frac{y_s}{D} - 0.01\right)\right] W_b \quad \text{if } 0.01 \le \frac{y_s}{D} \le 0.1 \tag{5-14}$$

where y_s is the mean depth of sediment over the pipe invert and D is the pipe diameter. If the criteria identified in Equation 5-14 is not met, then

$$W_e = \frac{W_b}{2} \tag{5-15}$$

The remaining coefficients of Equation 5-13 (i.e., J, K, m, α, β, λ, δ, and ε) can be determined using a dimensionless grain size, D_{gr}, defined as (Delleur, 2001)

$$D_{gr} = d\left[\frac{g(S-1)}{\nu^2}\right]^{1/3} \tag{5-16}$$

For $1 \leq D_{gr} \leq 60$ (i.e., finer sediment), the coefficients are given as

$$m = 1.67 + \frac{6.83}{D_{gr}} \tag{5-17}$$

$$n = 1 - 0.56 \log D_{gr} \tag{5-18}$$

$$J = \frac{8^{\frac{n(1-m)}{2}} H}{113^{m(1-n)} A_{gr}^{m}} \tag{5-19}$$

$$\log H = 2.79 \log D_{gr} - 0.98 (\log D_{gr})^2 - 3.46 \tag{5-20}$$

$$A_{gr} = 0.14 + \frac{0.23}{\sqrt{D_{gr}}} \tag{5-21}$$

$$\alpha = 1 - n \tag{5-22}$$

$$\beta = \frac{10 - 4m - mn}{10} \tag{5-23}$$

$$\lambda = \frac{n(m-1)}{2} \tag{5-24}$$

$$\delta = \frac{n}{2} \tag{5-25}$$

$$K = 11.3^{\alpha} g^{\delta} A_{gr} \tag{5-26}$$

$$\varepsilon = \frac{4+n}{10} \tag{5-27}$$

For $D_{gr} > 60$ (i.e., coarse sediment) $m = 1.78$, $n = 0$, $A_{gr} = 0.17$, and $H = 0.025$.

For movement of bed load at the limit of deposition (i.e., no deposition), CIRIA recommends the following expression that provided better overall performance when used with a number of independent sets of laboratory data (Ackers et al., 1996; Delleur, 2001)

$$C_v = 0.0303\left(\frac{D^2}{A}\right)\left(\frac{d}{D}\right)^{0.6}\left(1-\frac{V_T}{V}\right)^4\left(\frac{V^2}{g(S-1)D}\right)^{1.5} \tag{5-28}$$

where V_T is the flow velocity at the threshold of bed load movement at which individual particles begin to move along the invert of the pipe and is given as

$$V_T = 0.125\sqrt{g(S-1)d}\left(\frac{y}{D}\right)^{0.47} \tag{5-29}$$

in which y is flow depth. For the case of bed load with deposition, the following equation by May (1995) is recommended (Delleur, 2001)

$$C_v = \eta\left(\frac{W_b}{D}\right)\left(\frac{D^2}{A}\right)\left(\frac{\theta f_g V^2}{8g(S-1)D}\right) \tag{5-30}$$

where η is a transport parameter for a continuous bed; W_b is the sediment bed width; f_g is the Darcy-Weisbach friction factor corresponding to the grain shear stress; and θ_t is a transition coefficient for the particle Reynolds number. The transport parameter depends on a grain mobility factor, F_g, determined by (Delleur, 2001)

$$F_g = \left(\frac{f_g V^2}{8g(S-1)d}\right)^{1/2} \tag{5-31}$$

where f_g can be calculated using the Colebrook-White or similar equation with an equivalent sand grain roughness (i.e., k_s) of 1.23 times the median grain diameter. The particle Reynolds number and transition coefficient are evaluated as

$$Re^* = \left(\frac{f_c}{8}\right)^{1/2}\left(\frac{Vd}{\nu}\right) \tag{5-32}$$

and

$$\theta_t = \frac{e^{(Re^*/12.5)} - 1}{e^{(Re^*/12.5)} + 1} \tag{5-33}$$

The effective sediment mobility, F_s, is then

$$F_s = F_g\sqrt{\theta_t} \tag{5-34}$$

and the transport parameter, η, is determined according to Table 5-1.

Table 5-1: Transport parameter evaluation

F_s	η
$\leq$ 0.1	0
0.1 to 0.225	1.2 (F_s - 0.1)
0.225 to 0.275	0.15 + 9.0 (F_s - 0.225)
0.275 to 0.4	0.6 + 3.2 (F_s - 0.275)
0.4 to 0.7	1.0 - (F_s - 0.4)
0.7 to 0.8	0.7

Source: Adapted from Delleur (2001)

5.6 BIOCHEMICAL OXYGEN DEMAND

Biochemical oxygen demand (BOD) is the mass of oxygen used by indigenous microbes in one liter of water when they are subjected to biodegradable organics. Thus, it represents the amount of oxygen required to biochemically stabilize organic matter that is present in sewer flows. Discharges that contain significant amounts of organic matter have high BOD levels and consume large amounts of oxygen. In turn, the dissolved oxygen in receiving aquatic systems will be adversely lowered by these discharges and negatively impact aquatic ecosystems. The ability to model BOD is important in wastewater engineering to assist wastewater utilities in (1) estimating the quantity of oxygen required to biologically stabilize the organic matter present; (2) determining the size of wastewater treatment facilities; (3) evaluating the efficiency of the treatment process; and (4) ensuring compliance with discharge permits (Tchobanoglous, 2003).

The rate of BOD exertion changes over time due to the type of organisms present. Carbonaceous BOD is first exerted by heterotrophic organisms, whereas nitrogenous BOD is subsequently exerted by nitrifying bacteria. Normally, the latter is only significant after eight to ten days of oxidation in the presence of excess oxygen. As shown in Figure 5-1, the carbonaceous BOD will eventually reach an ultimate value (i.e., BOD_L), as will the nitrogenous component. It is common practice, however, to characterize gross amounts of organic matter using a five-day BOD (i.e., BOD_5), representing the oxygen required over a period of five days. For example, municipal wastewaters typically have BOD_5 levels ranging from 110 to 400 mg/L.

Using first order kinetics, the variation of BOD with time, t, is given as

$$BOD = BOD_L\left(1 - e^{-tk_1}\right) \tag{5-35}$$

where k_1 is a rate constant estimated from laboratory data and ranges from approximately 0.05 to 0.2 day^{-1}. Thus, the BOD_5 and BOD_L are related by

$$BOD_L = \frac{BOD_5}{1 - e^{-5k_1}} \tag{5-36}$$

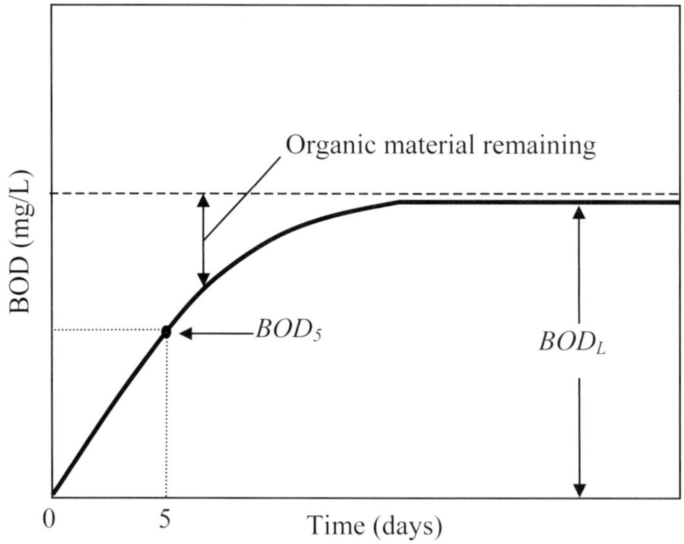

Figure 5-1: Time variation of carbonaceous BOD

5.7 *HYDROGEN SULFIDE*

The control of odorous gases and corrosion of sewer facilities are the two most important problems in operating wastewater collection systems. Both problems are mainly related to the generation of hydrogen sulfide. Evaluation of existing or potential odor and corrosion problems, and identification of where they will occur is, therefore, highly essential for sewer systems. Hydrogen sulfide (H_2S) has a characteristic rotten egg odor, is extremely toxic, and is a precursor to the formation of sulfuric acid (which corrodes concrete, lead-based paints, metals, and other materials). The conditions leading to the formation of H_2S generally favor the production of other odorous compounds. Therefore, investigation of the conditions favoring formation of H_2S also helps to quantify the potential for odor generation from other organic compounds.

In wastewater collection systems, odor is commonly produced as a result of biological decomposition of organic matter, particularly those containing sulfur and nitrogen, under anaerobic conditions prevailing in the slime layer of gravity pipes, force mains, and wet wells. Hydrogen sulfide and ammonia are the malodorous inorganic gases produced from the decomposition. However, H_2S is the most commonly known and prevalent odorous gas associated with domestic wastewater collection and treatment systems. Reduction of the sulfate ion, one of the most universal anions occurring in natural waters, under anaerobic conditions as shown by the following reaction, is the primary source of H_2S in wastewater collection systems.

$$SO_4^{=} + \text{organic matter} \xrightarrow[\text{bacteria}]{\text{anaerobic}} S^{=} + H_2O + CO_2$$

$$S^{=} + 2H^{+} \longrightarrow H_2S$$

The molecular H_2S, formed from reduction of sulfate ion, dissolves in the wastewater and dissociates in accordance with the reversible ionization reaction expressed as

$$H_2S \longleftrightarrow HS^{-} + H^{+}$$

$$HS^{-} \longleftrightarrow S^{=} + H^{+}$$

Partitioning of hydrogen sulfide into H_2S aqueous, HS^- ion, and $S^=$ ion depends primarily on the temperature and pH of the wastewater, although ionic strength, as represented by dissolved solids or electrical conductivity, also affects the dissociation. The major portion of sulfides generated and available in sewer systems is retained in solution as a mixture of hydrogen sulfide and ionic HS^-; the mixture is called dissolved sulfide. Table 5-2 shows proportions of dissolved sulfide present as H_2S at various pH levels. The table values are obtained assuming that the negative logarithm of the practical ionization constant in wastewater is 7.0. Some of the aqueous H_2S will escape into the sewer atmosphere causing odor problems. The HS^- ion and $S^=$ ion produce no odors. The rate of escape of H_2S gas into the sewer atmosphere is a function of the difference between the saturation or equilibrium concentration determined by Henry's law and the actual concentration of H_2S in the sewer atmosphere. Figure 5-2 shows H_2S in the sewer atmosphere in equilibrium with the given concentrations of aqueous H_2S concentration in the wastewater at the respective temperatures, for a pressure of one atmosphere (USEPA, 1985).

Table 5-2: Proportion of dissolved sulfide present as H_2S

pH	Percent H_2S
5.0	99
6.0	91
6.5	76
7.0	50
7.1	44
7.2	39
7.4	28
7.5	24
7.6	20
7.7	17
7.8	14
7.9	11
8.0	9

Source: Adapted from ASCE (1982)

The rate of sulfide buildup in wastewater collection systems depends on environmental, pipe, and wastewater conditions. Specifically, variables such as concentration of organic material and nutrients, sulfate concentration, dissolved oxygen (DO), pH, temperature, stream velocity, pipe slope, surface area of the flow, and detention (residence) time play crucial role. The potential for sulfide buildup in gravity sewers having a diameter of less than approximately 2 ft (0.6 m) can be assessed by (Pomeroy and Parkhurst, 1977)

$$Z = 0.308\frac{(EBOD)}{S_o^{0.5}Q^{0.33}}\left(\frac{P}{B}\right) \tag{5-37}$$

where

$$EBOD = BOD_5 \times 1.07^{T-20} \tag{5-38}$$

and where *EBOD* is an effective BOD, in mg/l; BOD_5 is in mg/l and taken during the highest six-hour flow period of the day; T is temperature of the wastewater in °C; Q is in m^3/s; P is wetted perimeter in m; and B is top width of sewer flow in m. When the value of Z is less than 5,000, sulfide is rarely generated. For $5{,}000 \le Z \le 10{,}000$, there exists a marginal condition for sulfide generation, and when Z exceeds 10,000, sulfide generation can be expected

(ASCE, 1982). Empirical equations that are used to predict the rate of sulfide buildup in sewer systems are presented below.

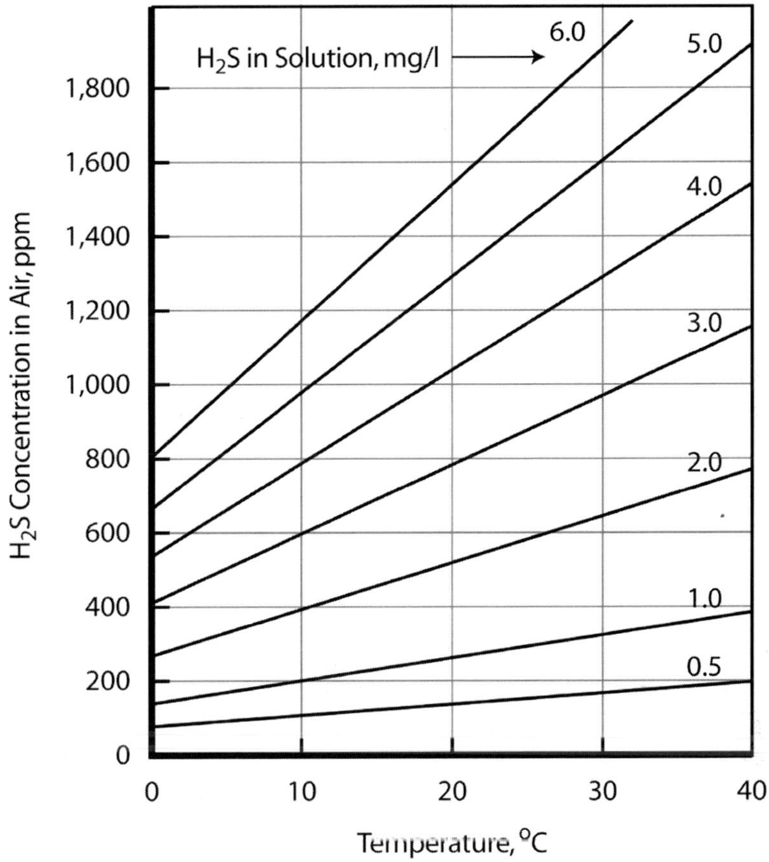

Source: Adapted from USEPA (1985)

Figure 5-2: Equilibrium concentration of H_2S in air

To predict buildup of sulfide in gravity sanitary sewer pipes flowing less than full, Pomeroy and Parkhurst (1977) presented the following equation based on research done on the Los Angeles County wastewater collection systems and some other data. The equation applies only to pipes with slope less than 0.6%, and in which dissolved oxygen is less than 0.5 mg/l.

$$\frac{d[S]}{dt} = \frac{M'EBOD}{R} - \frac{m[S](sV)^{3/8}}{D} \qquad (5\text{-}39)$$

where $d[S]/dt$ is the rate of change of total sulfide concentration, in mg/l-hr; M' is an effective sulfide flux coefficient for sulfide generation by the slime layer in gravity sewers, in m/hr; *EBOD* is an effective BOD obtained using Equation 5-38; T is wastewater temperature, in ^{0}C; R is hydraulic radius, in m; m is a dimensionless empirical coefficient to account for sulfide losses by oxidation and escape to atmosphere, $[S]$ is total sulfide concentration, in mg/l; D is hydraulic depth, in m; V is mean sewage velocity, in m/s; and s is slope of energy grade line. The coefficients M' and m could be calibrated to the condition of the system under consideration. However, two sets of assumed values, moderately conservative and more conservative, are commonly used for the coefficients. The moderately conservative condition uses 0.00032 m/hr and 0.96 for M' and m, respectively, and it represents systems with low dissolved oxygen where sulfide buildup is in progress. The more conservative assumption uses 0.00032 m/hr and 0.64 for M' and m, respectively.

At equilibrium conditions, $d[S]/dt = 0$. Solving for S, the theoretical upper limit at equilibrium, yields

$$S_{lim} = (M'/m)EBOD(sV)^{-3/8}(P/B) \tag{5-40}$$

where S_{lim} is the limiting sulfide concentration, in mg/l; P is wetted perimeter of the sewer pipe, in m; and B is surface width of the pipe, in m.

Sulfide generation at the downstream end of a pipe with uniform slope and flow can be predicted from

$$S_2 = S_{lim} - \frac{(S_{lim} - S_1)}{log^{-1}\left[\dfrac{m(sV)^{3/8}t}{2.31D}\right]} \tag{5-41}$$

where S_2 is sulfide concentration at the downstream end of the pipe, in mg/l; t is flow travel time in the sewer reach, in hr.

For pipes flowing full (e.g., force mains and surcharged pipes), sulfide buildup is estimated using

$$\frac{d[S]}{dt} = M\left[EBOD\left(\frac{4}{d} + 1.57\right)\right] \tag{5-42}$$

where d is pipe diameter, in m; M is sulfide flux coefficient for filled pipes, in m/hr. Value of 0.001 is commonly used for M. The remaining variables are as previously defined.

Integrating and solving for S_2, yields

$$S_2 = S_1 + (M)(t)\left[EBOD\left(\frac{4}{d} + 1.57\right)\right] \tag{5-43}$$

As described previously, H_2S is an acutely toxic material. It can cause serious health hazards even at very low concentrations. The physiological effects of H_2S gas are summarized in Figure 5-3.

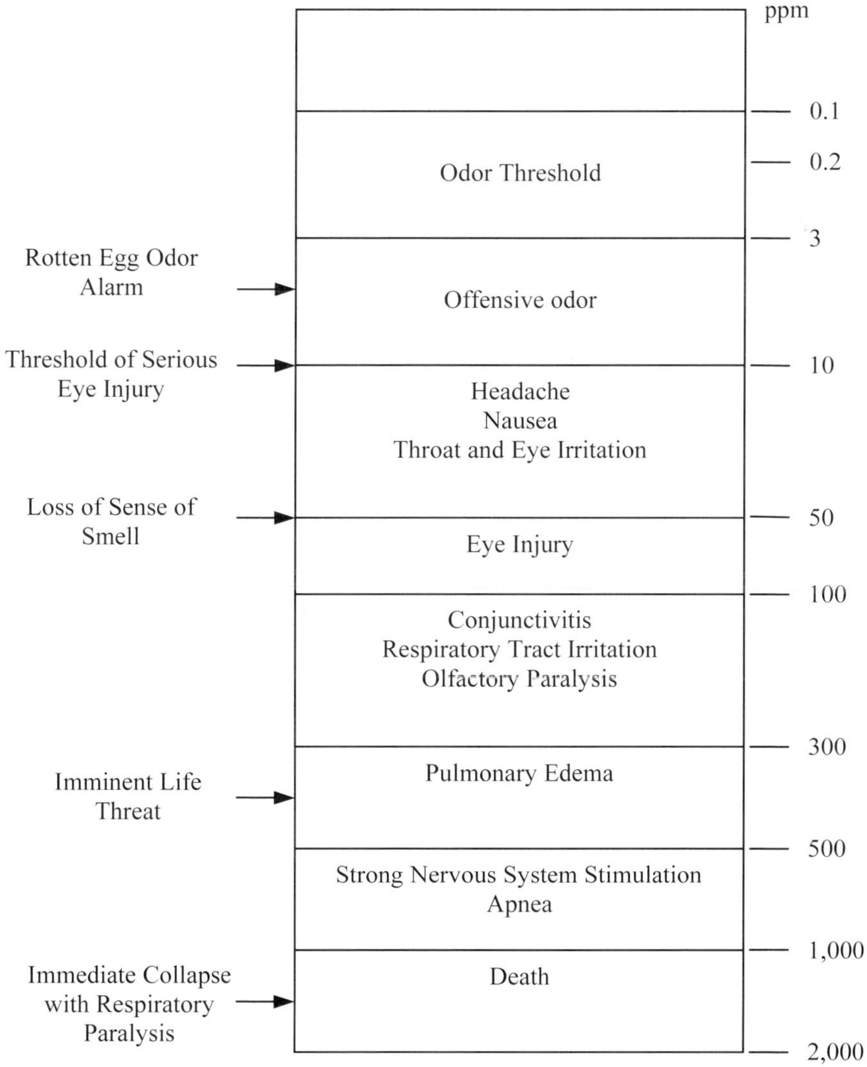

Source: Adapted from USEPA (1985)

Figure 5-3: Hydrogen sulfide toxicity spectrum

Overall, estimation of sulfide buildup and release can assist the wastewater utility to pinpoint odor problems, develop effective monitoring programs, alert plant operators and maintenance workers to potential danger and the need to observe safety practices, and evaluate and implement effective control systems such as aeration, chlorination, and mechanical cleaning.

5.8 CORROSION

Corrosion is one of the primary reasons that sewer systems lose their structural integrity. Corroded sewer pipes may allow greater inflow and infiltration into the collection system, further deteriorating reliability of the network by causing undesirable conditions such as surcharges and overflows, ultimately requiring premature replacement of the pipes. Internal corrosion of unprotected concrete or metal surfaces is primarily due to the production of sulfuric acid in sewers through oxidation of hydrogen sulfide gas by bacterial action on the exposed surfaces under aerobic conditions. This type of corrosion is commonly referred to as microbially induced corrosion.

Some of the soluble H_2S available in partially filled pipes may escape into the sewer atmosphere and transferred to the pipe walls above the wastewater surface. It will then be taken up as it comes in contact with the damp surfaces of pipes. The H_2S retained in this dampness is then converted to sulfuric acid by aerobic bacteria, as described in the following reaction

$$H_2S + 2O_2 \xrightarrow{\text{bacteria}} H_2SO_4$$

Next, some or all of the produced sulfuric acid reacts with the pipe material causing corrosion, mainly at the inside pipe wall above the wastewater flow line. The amount of sulfuric acid that reacts with the pipe material depends on the rate of production of the acid, which in turn depends on moisture, the presence of oxygen, and emission of sulfide gas. In addition to concentration of acid present, the corrosive effect of sulfuric acid varies according to the type of pipe material used and the ambient temperature. Cementitious pipes, including ferrous pipes with mortar lining, experience a reaction that converts the surface material into a pasty mass, which is primarily a calcium sulfate, commonly referred to as gypsum. This mass may fall away from pipe walls and expose new surfaces to corrosive attack. Ferrous pipe materials may experience surface reaction in which a portion of the material is dissolved and converted to iron sulfide, yielding a hard bulky mass that forms on the exposed surface. A warm and humid environment creates a favorable condition for corrosion (Metcalf & Eddy, 1981).

In summary, the rate of microbially induced pipe corrosion depends upon the rate of sulfuric acid production, the amount of the produced sulfuric acid that reacts with the pipe material, and the alkalinity of the pipe material. Sulfuric acid production is related to the amount of hydrogen sulfide gas that escapes to the sewer atmosphere, which in turn depends on a number of environmental, pipe, and wastewater conditions. The empirical equations given below are commonly used to estimate the rate of microbially induced corrosion for cementitious (cement-bonded) materials and ferrous materials in partially filled sewers. For force mains, where the lines flow full, there is generally no internal corrosion since generation of sulfuric acid is prevented.

For cement-bonded materials

$$C = 11.4k\phi_{sw}\frac{1}{A} \tag{5-44}$$

where C is average rate of penetration, mm/yr; k is coefficient of efficiency for acid reaction considering the estimated fraction of acid remaining on the wall. The value could be as low as 0.3 and approaches 1.0 for complete acid reaction. ϕ_{sw} is flux of H_2S to the pipe wall, $g/m^2.hr$; A *is* alkalinity of the cement-bonded material, expressed as $CaCO_3$ equivalents. It takes value of 0.18 to 0.23 for granitic aggregate concrete, 0.9 for calcareous aggregate, 0.4 for mortar linings, and 0.5 for asbestos cement.

For ferrous materials

$$C = 2.04k\phi_{sw} \tag{5-45}$$

where *C, k,* and ϕ_{sw} are consistent with the definitions given above.

Assuming that all the H_2S escaping to the sewer atmosphere is oxidized on the pipe wall, average sulfide flux to the wall, ϕ_{sw}, can be estimated by multiplying the sulfide flux to the sewer atmosphere by the ratio of top (surface) width to exposed pipe wall perimeter, as

$$\phi_{sw} = 0.69(sV)^{3/8}\, j[DS](B/P') \tag{5-46}$$

where j is the proportion of dissolved sulfide present as H_2S (given in Table 5-2); *[DS]* is dissolved sulfide concentration in the wastewater, in mg/l; s is slope of energy grade line; and V is mean sewage velocity, in m/s; B is width of the wastewater flow at the surface, in m; and P' is exposed perimeter of the pipe wall above the water surface, in m. Equation 5-46 may not be applicable for sewer pipes that have shallow depth and high flow velocity, or for reaches with any points of high turbulence.

The capability to predict the rate of microbially induced corrosion in sewer collection systems under varying environmental and hydraulic conditions, aided by the empirical equations presented above, enables wastewater utilities to more effectively pinpoint facilities with potential corrosion problems, prioritize repairs, specify corrosion resistant materials or select other forms of corrosion protection measures (e.g., protective linings).

5.9 SOLVED PROBLEMS

Problem 5.1 Flow age

Flow travels through a 1,000-ft long storm sewer at 8 fps. Estimate the age of flow using the upstream end of the main as a reference.

Solution

From Equation 5-1, the age amounts to sewer length divided by velocity, or

$$T_j = 0 + \frac{1000}{8}\left(\frac{1\ min}{60\ sec}\right) = \underline{\underline{2.08\ min}}$$

Problem 5.2 Mass balance

A storm sewer junction collects runoff from two submains. Submain *A* carries 57 cfs with a suspended solids concentration of 90 mg/L, and submain *B* is estimated to carry 12 cfs. If regulations limit the suspended solids concentration that is distributed to a nearby receiving stream to 60 mg/L, determine the maximum concentration in the remaining submain.

Solution

Application of Equation 5-3 for the given data yields

$$80 = \frac{(57 \times 90) + (12 \times C)}{57 + 12}$$

Solving for *C*, the maximum suspended solids concentration is $\underline{\underline{32.5\ mg/L}}$.

Problem 5.3 Loadograph conversion

For the following upstream discharge hydrograph and pollutograph, compute the resulting loadograph:

Time (hr)	0	1	2	3	4	5	6	7	8
Discharge (cfs)	6	60	136	113	84	63	44	23	60
Concentration (mg/L)	2	36	32	25	18	15	12	8	5

Solution

For the first time interval, load is computed using Equation 5-4 as follows:

$$L = K_p CQ = (5.39)(2)(6)\left(\frac{1\ day}{24\ hrs}\right) = 2.70\ lb/hr$$

The entire loadograph is computed in a similar manner and is summarized in Table P5-3.

Table P5-3

(1)	(2)	(3)	(4)
Time (hr)	Discharge (cfs)	Concentration (mg/L)	Pollutant Load (lb/hr)
0	14	32	**100.6**
1	60	36	**485.1**
2	136	32	**977.4**
3	113	25	**634.5**
4	84	18	**339.6**
5	63	15	**212.2**
6	44	12	**118.6**
7	23	8	**41.3**
8	20	5	**67.4**

Problem 5.4 Muskingum method for conservative transport

Using the Muskingum method, route the hydrograph and loadograph from Problem 5.3 through a sewer reach. The pollutant under consideration is conservative. Assume $C_1 = 0.13$, $C_2 = 0.34$, and $C_3 = 0.53$ and that the initial discharge and load at the downstream end are 6 cfs and 34 lb/hr.

Solution

The Muskingum routing equation (see Chapter 3) for this application can be written in a slightly modified form. Flow can be expressed as

$$Q_D^{i+1} = C_1 Q_U^{i+1} + C_2 Q_U^{i} + C_3 Q_D^{i}$$

where subscripts D and U refer to downstream and upstream sections of the sewer reach, respectively, and superscript i refers to time increment. Thus, the discharge at 1 hr (i.e., $i = 0$) is computed by

$$Q_D^{i+1} = (0.13)(60) + (0.34)(14) + (0.53)(6) = 15.7 \ cfs$$

Similarly, the routing equation for the loadograph can be expressed as

$$L_D^{i+1} = C_1 L_U^{i+1} + C_2 L_U^{i} + C_3 L_D^{i}$$

so that at 1 hr

$$L_D^{i+1} = (0.13)(485.10) + (0.34)(100.61) + (0.53)(34) = 115.3 \ lb/hr$$

Columns (4) and (5) of Table P5-4 summarize the determination of the hydrograph and loadograph at the downstream end of the corresponding reach for remaining time increments.

Table P5-4

(1)	(2)	(3)	(4)	(5)
Time (hr)	Upstream		Downstream	
	Discharge (cfs)	Pollutant Load (lb/hr)	Discharge (cfs)	Pollutant Load (lb/hr)
0	14	100.6	**15.7**	**115.3**
1	60	485.1	**46.4**	**353.1**
2	136	977.4	**85.5**	**601.9**
3	113	634.5	**94.7**	**578.9**
4	84	339.6	**86.9**	**449.9**
5	63	212.2	**73.2**	**326.0**
6	44	118.6	**56.8**	**218.5**
7	23	41.3	**40.5**	**138.6**
8	20	67.4	**28.3**	**96.4**

Problem 5.5 Universal soil loss equation

The surface of a 5-ac construction site is comprised of bare sand. The overland flow length is 1,000 ft on an average slope of 0.001 ft/ft. If no erosion control measures are used, assess and comment on the suspended solids load generated at the site in response to a 3-hr rainfall event having the following intensities:

Time (hr)	0.0	0.5	1.0	1.5	2.0	2.5
Intensity (in/hr)	0.3	0.6	1.1	0.4	0.2	0.2

Solution

From the intensities provided, i_{30} is 1.1 in/hr. Applying Equation 5-6 for the first 0.5-hr interval yields

$$R_r = (1.1)[9.16 + 3.31 \log(0.3)](0.3)(0.5) = 1.23$$

The value of K for sand is assumed to be a constant 0.10. The dimensionless length-slope factor is evaluated as

$$L_S = \sqrt{1000}\left(0.0076 + (0.53)(0.001) + (0.76)(0.001)^2\right) = 0.26$$

Since the site consists of bare soil with no erosion control measures, both P_E and C can be taken as 1.0. The suspended solids load is computed using the USLE, Equation 5-5. For example, at the first time interval

$$E = (2000)(5)(1.23)(0.10)(0.26)(1.0)(1.0) = 319 \ lbs$$

Evaluations of R_r and E for remaining time intervals are shown in Table P5-5.

Note that values for A, K, L_s, C, and P_E remain constant for each time period. Summing the results in Column (4) yields a total load of 3,314 lbs. Even though results are only approximate, the load estimate is significant for this bare site and erosion control measures are well warranted.

Problem 5.6 Universal soil loss equation

If the site described in Problem 5.5 was covered by grass, assess the resulting change in estimated solids generated.

Solution

For grass, the land cover and management factor drops from 1.0 to approximately 0.003. Thus, each of the values in Column (4) of Table P5-5 would be multiplied by the new factor; alternatively, using the results from Problem 5.5, the total solids can be estimated as 3,314 lbs × 0.003, or 9.9 lbs. The solids generated can therefore be reduced by more than two orders of magnitude through influence of land cover and management practices implemented. Fortunately, this parameter, along with P_E, are variables that engineers and planners have the most influence over in trying to limit pollutant loads.

Table P5-5

(1)	(2)	(3)	(4)
Time Interval (hr)	Intensity (in/hr)	R_r	E (lbs)
0.00	0.3	1.23	**319**
0.50	0.6	2.78	**723**
1.00	1.1	5.62	**1,462**
1.50	0.4	1.73	**449**
2.00	0.2	0.75	**196**
2.50	0.2	0.75	**196**

Problem 5.7 Solids wash off

Prior to rainfall, a 10-ac impervious watershed has accumulated 3,200 lbs of solids. If the storm produces the runoff hydrograph listed below, determine the time distribution of solids washed off from the basin. Assume $k = 4.6$ in-1

Time (hr)	0.0	0.5	1.0	1.5	2.0	2.5	3.0	3.5	4.0
Discharge (cfs)	0	3	5	9	12	10	8	7	6

Solution

The rainfall depths are computed by multiplying the average runoff rate for each time increment by the corresponding length of that increment (i.e., 0.5 hr). For example, at $t = 0$

$$r = \left(\frac{0+3}{2}\right)(0.5)\frac{3600\ \ s/hr \times 12\ \ in/ft}{10\ \ ac \times 43{,}560\ \ ft^2 / ac} = 0.074\ \ in$$

Then from Equation 5-8, noting that the initial solids build up is 3,200 lbs, the amount remaining on the surface is expressed as

$$E = 3200 \times e^{-(4.6)(0.074)} = 2{,}276.8\ \ lbs$$

The solids washed off during the first 0.5 hr increment is therefore 3,200 – 2,276.8 = 923.2 lbs. Computations for the remaining time periods are shown in Table P5-7.

Table P5-7

(1)	(2)	(3)	(4)	(5)
Time Interval (hr)	Discharge (cfs)	r (in)	E (lbs)	ΔE (lbs)
0.0	0	0.074	**2,276.8**	**923.2**
0.5	3	0.198	**915.7**	**1,361.0**
1.0	5	0.347	**185.6**	**730.1**
1.5	9	0.521	**16.9**	**168.7**
2.0	12	0.546	**1.4**	**15.5**
2.5	10	0.446	**0.2**	**1.2**
3.0	8	0.372	**0.0**	**0.1**
3.5	7	0.322	**0.0**	**0.0**
4.0	6	0.149	**0.0**	**0.0**

Problem 5.8 Suspended load transport

Determine the velocity necessary to avoid deposition of suspended sediment in a 150-mm diameter pipe laid on a slope of 0.008 m/m. The suspended load is estimated to be 90 ppm. The material has a median grain diameter of 0.5 mm and a specific gravity of 2.65, while the pipe is characterized using a Manning roughness of 0.013 and carries approximately 0.010 m^3/s. Assume a composite friction factor of 0.025 and a particle fall velocity of 0.009 m/s.

Solution

Using Manning's equation (see Chapter 3), the central angle of flow, θ, is computed by

$$Q = \frac{K_m}{n} AR^{2/3} S_o^{\ 1/2} =$$

$$\frac{1.0}{0.013}\left[\frac{(\theta - \sin\theta)(0.15)^2}{8}\right]\left[\frac{0.15}{4}\left(1 - \frac{\sin\theta}{\theta}\right)\right]^{2/3}(0.008)^{1/2} = 0.010$$

Solution by trial-and-error, numerical analysis, or computational solver yields θ = 3.709 rad. From information in Table 3-1, depth y is equal to 0.096 m and

$$A = \frac{(\theta - \sin\theta)D^2}{8} = \frac{(3.709 - \sin 3.709)(0.15)^2}{8} = 0.012\, m^2$$

and

$$R=\frac{D}{4}\left(1-\frac{\sin\theta}{\theta}\right)=\frac{0.15}{4}\left(1-\frac{\sin 3.709}{3.709}\right)=0.043\ m$$

The shear stress can be approximated by

$$\tau=(9810)(0.043)(0.008)=3.37\ N/m^2$$

which exceeds the criteria necessary for application of Equation 5-9. Thus, using the data provided, the Macke formula can be expressed as

$$C_v=\frac{(0.025)^3 V_L^3}{30.4(2.65-1)(0.009)^{1.5}0.012}=9\times 10^{-5}$$

Solving for the V_L yields a limiting velocity of 0.14 m/s.

Problem 5.9 Suspended load transport

For the data given in Problem 5.8, determine the suspended sediment concentration if deposition occurs and the mean depth of sediment is 7.5 mm.

Solution

The velocity is computed by

$$V=\frac{Q}{A}=\frac{0.01}{0.012}=0.83\ m/s$$

Application of Equation 5-13 first requires determination of a number of parameters and coefficients. The dimensionless grain size is evaluated as

$$D_{gr}=\left(0.5\times 10^{-3}\right)\left[\frac{9.81(2.65-1)}{\left(1\times 10^{-6}\right)^2}\right]^{1/3}=12.65$$

Then from application of Equations 5-17 through 5-27, the following values are determined: $m = 2.21$, $n = 0.38$, $H = 0.027$, $A_{gr} = 0.20$, $J = 9 \times 10^{-4}$, $\beta = 0.032$, $\alpha = 0.62$, $\lambda = 0.23$, $\delta = 0.19$, $K = 1.39$, and $\varepsilon = 0.44$. The relationship between bed width (i.e., sediment top width) and depth can be obtained from the same relationships presented for flow in Table 3-1, or

$$W_b=2\sqrt{y_s(D-y_s)}=2\sqrt{(0.0075)(0.15-0.0075)}=0.065\ m$$

Since $y_s/D = 7.5/150 = 0.05$, Equation 5-14 applies, so that

$$W_e=[0.2+3.33(0.05-0.01)]0.065=0.022\ m$$

and Equation 5-13 gives

$$C_v = 9\times10^{-4}\left[\frac{(0.022)(0.043)}{0.012}\right]^{0.62}\left[\frac{0.5\times10^{-3}}{0.043}\right]^{0.032} 0.025^{0.23}\times$$

$$\left[\frac{0.83}{\sqrt{9.81(2.65-1)0.043}}-(1.39)(0.025)^{0.19}\left(\frac{0.5\times10^{-3}}{0.043}\right)^{0.44}\right]^{2.21} = 5.5\times10^{-5}$$

or 55 ppm by volume.

Problem 5.10 Bed load transport

Assuming that no deposition is permitted, determine the maximum bed sediment loading rate for a 150-mm pipe having a Manning roughness of 0.013 and laid on a slope of 0.009 m/m. The material is characterized as having a median grain diameter of 0.78 mm and a specific gravity of 2.65, and the flow rate is estimated to be 0.011 m^3/s.

Solution

Using Manning's equation, the central angle of flow, θ, is computed by

$$Q = \frac{K_m}{n} AR^{2/3} S_o^{1/2} =$$

$$\frac{1.0}{0.013}\left[\frac{(\theta-\sin\theta)(0.15)^2}{8}\right]\left[\frac{0.15}{4}\left(1-\frac{\sin\theta}{\theta}\right)\right]^{2/3}(0.009)^{1/2} = 0.011$$

Solution by trial-and-error, numerical analysis, or computational solver yields θ = 3.765 rad. The corresponding depth y is 0.098 m and

$$A = \frac{(\theta-\sin\theta)D^2}{8} = \frac{(3.765-\sin 3.765)(0.15)^2}{8} = 0.012\,m^2$$

and

$$V = \frac{Q}{A} = \frac{0.011}{0.0122} = 0.90\,m/s$$

The threshold velocity is computed as follows using Equation 5-29:

$$V_T = 0.125\sqrt{9.81(2.65-1)(0.78\times10^{-3})}\left(\frac{0.098}{0.15}\right)^{0.47} = 0.012\,m/s$$

The sediment concentration is then approximated as

$$C_v = 0.0303\left(\frac{0.15^2}{0.012}\right)\left(\frac{0.78\times10^{-3}}{0.15}\right)^{0.6}\times$$

$$\left(1-\frac{0.012}{0.90}\right)^4\left(\frac{0.90^2}{9.81(2.65-1)0.15}\right)^{1.5} = 4.4\times10^{-4}$$

or <u>440 ppm</u> by volume.

Problem 5.11 Bed load transport

For the data provided in Problem 5.10, estimate the bed load concentration if deposition occurs. Assume that the bed sediment width across the pipe invert has reached 80 mm.

Solution

The Darcy-Weisbach friction factor for grain shear is computed using the Colebrook-White equation and $k_s = 1.23 \times 0.78$ mm, or 0.96 mm. The hydraulic radius and Reynolds number are approximately

$$R = \frac{D}{4}\left(1-\frac{\sin\theta}{\theta}\right) = \frac{0.15}{4}\left(1-\frac{\sin 3.765}{3.765}\right) = 0.043\ m$$

$$Re = \frac{V4R}{\nu} = \frac{(0.90)(4\times0.043)}{1\times10^{-6}} = 154{,}800$$

Then,

$$\frac{1}{\sqrt{f}} = -2\log\left(\frac{0.96\times10^{-3}}{14.83(0.043)} + \frac{2.52}{Re\sqrt{f}}\right)$$

Solving iteratively yields $f = f_g = 0.032$. The grain mobility factor is computed from Equation 5-31 as

$$F_g = \left(\frac{(0.032)(0.90)^2}{8(9.81)(2.65-1)(0.78\times10^{-3})}\right)^{1/2} = 0.51$$

For a clean pipe, the same approach used previously can be used to evaluate f_w. Assuming an equivalent sand grain roughness of 0.025 mm for the pipe, then f_w = 0.017. In addition, the wetted perimeter of the pipe wall can be approximated as

$$P_w = \frac{A}{R} - W_b = \frac{0.012}{0.043} - 0.080 = 0.20\ m$$

so that the composite Darcy-Weisbach friction factor can be computed via Equation 5-10 as

$$f_c = \frac{(0.20)(0.017) + (0.08)(0.032)}{0.20 + 0.08} = 0.021$$

The particle Reynolds number and θ_t are evaluated from Equation 5-32 and 5-33 as

$$Re^* = \left(\frac{0.021}{8}\right)^{1/2}\left(\frac{(0.90)(0.78 \times 10^{-3})}{1 \times 10^{-6}}\right) = 36$$

$$\theta_t = \frac{e^{(36/12.5)} - 1}{e^{(36/12.5)} + 1} = 0.15$$

The effective sediment mobility from Equation 5-34 and corresponding transport parameter (i.e., Table 5-1) are

$$F_s = 0.51\sqrt{0.15} = 0.2$$

$$\eta = 1.2(0.2 - 0.1) = 0.12$$

Then application of Equation 5-30 for sediment concentration gives

$$C_v = 0.12\left(\frac{0.080}{0.15}\right)\left(\frac{0.15^2}{0.012}\right)\left(\frac{(0.15)(0.032)(0.90)^2}{8(9.81)(2.65 - 1)(0.15)}\right) = 2.40 \times 10^{-5}$$

or 24 ppm by volume.

Problem 5.12 Biochemical oxygen demand

Laboratory tests on a sample of wastewater indicate a BOD_5 of 85 mg/L at 20°C and a rate constant, k_1, equal to 0.2 d^{-1}. Determine the ultimate carbonaceous BOD to be exerted (i.e., BOD_L).

Solution

BOD_L can be expressed directly using Equation 5-36.

$$BOD_L = \frac{85}{1 - e^{-5(0.2)}} = 134.5\ mg/L$$

Problem 5.13 Biochemical oxygen demand

For the data in Problem 5.12, estimate the time required for 95 percent of the carbonaceous BOD to be exerted.

Solution

From Equation 5-35, assuming BOD is equal to 0.95 × BOD_L,

$$0.95(134.5) = 134.5\left(1 - e^{-t(0.2)}\right)$$

which yields $t =$ 15.0 days.

Problem 5.14 Biochemical oxygen demand

If the ultimate nitrogenous BOD of the sample described in Problem 5.12 is equivalent to 15 percent of the carbonaceous BOD_L, determine the oxygen consumed per m^3 of wastewater.

Solution

The ultimate nitrogenous BOD is 1.15 × BOD_L, or 154.7 mg/L. Thus the mass of oxygen consumed is computed as

$$154.7\left(10^3\ L/m^3 \times 10^{-3}\ g/mg\right) = \underline{\underline{154.7\ g/m^3}}$$

Problem 5.15 H_2S generation

The sample from Problem 5.12 was taken from a 1-m diameter concrete pipe (n = 0.013) laid on a slope of 0.007 m/m and carrying 1.5 m^3/s. Evaluate the potential for sulfide generation if the wastewater temperature is 35° C.

Solution

Using Manning's equation, the central angle of flow is found by

$$Q = \frac{K_m}{n} AR^{2/3} S_o^{1/2} =$$

$$\frac{1.0}{0.013}\left[\frac{(\theta - \sin\theta)(1)^2}{8}\right]\left[\frac{1}{4}\left(1 - \frac{\sin\theta}{\theta}\right)\right]^{2/3}(0.007)^{1/2} = 1.5$$

which yields $\theta = 3.722$ rad. Then from Table 3-1,

$$P = \frac{\theta D}{2} = \frac{(3.722)(1)}{2} = 1.86\ m$$

and

$$B = D\,sin\frac{\theta}{2} = (1)sin\frac{3.722}{2} = 0.96\ m$$

From Equation 5-37

$$Z = 0.308\frac{\left(85 \times 1.07^{35-20}\right)}{(0.007)^{0.5}(1.5)^{0.33}}\left(\frac{1.86}{0.96}\right) = 1.5 \times 10^{3}$$

The computed index value indicates that H_2S will rarely be generated.

Problem 5.16 H_2S generation

Estimate the amount of dissolved sulfide at the downstream end of a 1000-m long, 1-m diameter sanitary sewer pipe. The pipe has been designed to carry the design sewage flow of 0.41 m^3/sec at flow depth to pipe diameter ratio of 0.5 with slope of 0.001. Manning's n for the pipe is 0.012. Assume the following characteristics for the wastewater: BOD_5 = 300 mg/l; insoluble sulfide = 0.1 mg/l; total sulfide at the upstream end of the pipe = 1.2 mg/l; temperature = 30^0C; and pH = 7.2. Use more conservative values for *M'* and *m* (i.e., *M'* = 0.00032 m/hr and *m* = 0.64).

Solution

First, it is essential to check if the given conditions favor sulfide generation.

Effective BOD of the wastewater is calculated using Equation 5-38 as

$$EBOD = 300 \times 1.07^{(30-20)} = 590.15\ mg/l$$

Since the sewer pipe is designed to discharge the sewage flow at half-full (i.e. depth/diameter = 0.5), surface width of the pipe is equal to diameter of the pipe (i.e., 1-m). Likewise, wetted perimeter of the pipe is

$$P = \frac{\pi \times diameter}{2} = 3.14 \times 0.5 = 1.57\ m$$

From Equation 5-37,

$$Z = 0.308\frac{(590.15)}{(0.001)^{0.5}(0.41)^{0.33}}\left(\frac{1.57}{1}\right) = 6{,}724$$

The *Z* index indicates existence of marginal conditions for sulfide generation. Therefore, it is essential to predict the amount of sulfide buildup. Since the pipe is designed to flow at half-full condition, the empirical equations developed to predict sulfide buildup in partially filled sanitary sewer pipes would be used to solve this specific problem.

Wetted area of the pipe is

$$Area = \frac{\pi \times diameter^2}{8} = \frac{3.14}{8} = 0.3925\, m^2$$

Hydraulic radius, *R*, and the hydraulic depth, *D*, are

$$R = \frac{0.3925}{1.57} = 0.25\, m\text{; and } D = \frac{0.3925}{1} = 1.0\, m$$

From Manning's equation, mean velocity of the sewage flow is

$$V = \frac{1}{0.012} 0.25^{2/3} 0.001^{1/2} = 1.046\, m/s$$

From Equation 5-40, the limiting sulfide concentration is

$$S_{lim} = \left(\frac{0.00032}{0.64}\right) \times 590.15 \times (0.001 \times 1.046)^{-3/8} \left(\frac{1.57}{1}\right)$$

$$= 6.05 \text{ mg/l}$$

Travel time of the sewage through the sewer pipe can be estimated by dividing length of the pipe with flow velocity

$$t = \frac{1000}{1.046} = 956.0\, sec = 0.266 \text{ hrs}$$

From Equation 5-41, total sulfide at downstream end of the sewer pipe is

$$S_2 = 6.05 - \frac{(6.05 - 1.2)}{log^{-1}\left[\frac{0.64(0.001 \times 1.046)^{3/8} \times 0.266}{2.31 \times 0.3925}\right]} = 1.36 \text{ mg/l}$$

Since the sewer pipe has 0.1 mg/l of insoluble sulfides, total dissolved sulfide at downstream end of the reach will be $S_2 - 0.1 = \underline{\underline{1.26\, mg/l}}$

Problem 5.17 H_2S generation

For the wastewater conditions given in Problem 5.16, assume that the pipe is a 2000-m long force main of 12-in diameter. Estimate the amount of dissolved sulfide at downstream end of the force main. Use 0.001 m/hr for sulfide flux coefficient.

Solution

Force mains flow under full flow conditions. Therefore, the empirical equations developed to predict sulfide buildup in pipes flowing full will be used to solve this problem. Velocity of flow through the force main is

$$V = \frac{Q}{Area} = \frac{0.41}{\left(\frac{\pi(0.3048)^2}{4}\right)} = 5.62\, m/s$$

Travel time of the sewage through the force main is

$$t = \frac{2000}{5.62} = 355.75\, sec = 0.1 \text{ hrs}$$

From Equation 5-43, total sulfide at downstream end of the force main is

$$S_2 = S_1 + (M)(t)\left[EBOD\left(\frac{4}{diameter} + 1.57\right)\right]$$

$$= 1.2 + 0.001 \times 0.1\left[590.15\left(\frac{4}{0.3048} + 1.57\right)\right] = 2.06\, mg/l$$

Therefore, total dissolved sulfide at downstream end of the force main will be

$$S_2 - 0.1 = 2.06 - 0.1 = \underline{1.96\, mg/l}$$

Problem 5.18 Corrosion rate prediction

For the pipe and wastewater conditions given in Problem 5.16, estimate the annual rate of microbially induced corrosion at the downstream end of the sewer pipe. Assume that the pipe is made of concrete with a granitic aggregate, and use alkalinity of 0.2. Assume 0.8 for acid reaction efficiency coefficient.

Solution

From Problem 5.16, we know that pH of the wastewater is 7.2. At this pH level, from Table 5-2, the percentage of dissolved sulfide that exists in the form of H_2S is 39. Since the pipe is flowing half-full, exposed perimeter of the pipe (P') is equal to wetted perimeter of the pipe (i.e., 1.57 m).

Most of the information needed to predict the rate of corrosion in the sewer pipe, including flow velocity and the concentration of dissolved sulfide at the downstream end of the pipe, is obtained while solving Problem 5.16. Therefore, flux of H_2S to the pipe wall could be estimated using Equation 5-46 as

$$\phi_{sw} = 0.69(sV)^{3/8} j[DS](B/P')$$
$$= 0.69(0.001 \times 1.046)^{3/8}(0.39)(1.158)(1/1.57)$$
$$= 0.015\ g/m^2\text{-}hr$$

Form Equation 5-44, the average rate of annual corrosion penetration would be

$$C = 11.4 \times 0.8 \times 0.015\left(\frac{1}{0.2}\right) = \underline{\underline{0.684\ mm/yr}}$$

REFERENCES CITED

Ackers, P., "Sediment Aspects of Drainage and Outfall Design," *Proc. Int. Conf. on Environ. Hydraulics*, Hong Kong, A.A. Balkema, Rotterdam, 215-230, 1991.

Ackers, J.C., D. Butler, and R.W.P. May, *Design of Sewers to Control Sediment Problems*, Construction Industry Research and Information Association (CIRIA), Westminster, London, 1996.

American Society of Civil Engineers (ASCE), *Gravity Sanitary Sewer Design and Construction*, ASCE Manuals and Reports on Engineering Practice No. 60 and Water Pollut. Control Fed. Manual of Practice FD-5, New York, NY, 1982.

Boulos, P.F., K.E. Lansey, and B.W. Karney, *Comprehensive Water Distribution Systems Analysis Handbook for Engineers and Planners*, MWH Soft, Inc., Pasadena, CA, 2004.

Butler, D., R. May, and J. Ackers, "Self-Cleansing Sewer Design Based on Sediment Transport Principles," *J. of Hydraulic Engrg*., vol. 129, no. 4, 276-282, 2003.

DeSutter, R., P.Rushforth, S. Tait, M. Huygens, R. Verhoeven, and A. Saul, "Validation of Existing Bed Load Transport Formulas Using In-Sewer Sediment," *J. of Hydraulic Engrg*., vol. 129, no. 4, 325-333, 2003.

Delleur, J.W., "Sediment Movement in Drainage Systems," in *Stormwater Collection Systems Design Handbook*, ed. by L.W. Mays, McGraw Hill, New York, NY, 2001.

Delleur, J.W., "Hydraulics of Sediment Movement in Urban Drainage Systems," *J. of Hydraulic Engrg*., vol. 129, no. 4, 251-252, 2003.

Huber, W.C., and R.E. Dickinson, *Storm Water Management Model, User's Manual*, Environmental Research Laboratory, Office of Research and Development, U.S. Environmental Protection Agency, Athens, GA, 1988.

May, R.W.P., *Sediment Transport Equation for Self-Cleansing Sewers*, HR Wallingford, Report SR 436, 1995.

Metcalf and Eddy, Inc., *Wastewater Engineering: Treatment, Disposal, and Reuse*, 3rd ed., McGraw-Hill, New York, NY, 1991.

Metcalf and Eddy, Inc., *Wastewater Engineering: Collection and Pumping of Wastewater*, McGraw-Hill, New York, NY, 1981.

McCuen, R.H., *Hydrologic Analysis and Design*, 3rd ed., Prentice-Hall, Upper Saddle River, NJ, 2005.

Pomeroy, R.D., and J.D. Parkhurst, "The Forecasting of Sulfide Buildup Rates in Sewers," *Progress in Water Technology*, vol. 9, 1977.

Sonnen, M.B., "Urban Runoff Quality: Information Needs," *J. of Tech. Councils*, ASCE, vol. 106(TCI), 29-40, 1980.

Sartor, J.D., and G.B. Boyd, "Water Pollution Aspects of Street Surface Contaminants," *Rep. EPA-R-272-081*, U.S. Environmental Protection Agency, Washington DC, 1972.

Tchobanoglous, G., *Wastewater Engineering Treatment and Reuse*, McGraw-Hill, New York, NY, 2003.

U.S. Environmental Protection Agency (USEPA), "Design Manual: Odor and Corrosion Control in Sanitary Sewerage Systems and Treatment Plants," *Rep. EPA/625/1-85/018,* U.S. Environmental Protection Agency, Cincinnati, OH

Wischeimer, W.H., and D.D. Smith, "Rainfall Energy and its Relationship to Soil Loss," *Trans. of Am. Geophysical Union*, vol. 39, no. 2, 285-291, 1958.

CHAPTER SIX

Management of Storm and Combined Sewer Overflows

6.1 INTRODUCTION

Various retardation structures exist for managing the increased volume of storm and combined sewer flows that are associated with urban development. While the most common of these are surface detention basins, other facilities include retention basins, infiltration basins, and subsurface detention systems. Each of these is designed to reduce the frequency of high, potentially erosive, flows and mitigate their adverse impacts. In addition, these structures provide water quality benefits in the form of pollutant removal and, as a result, are often referred to as structural best management practices (BMPs).

6.2 SURFACE DETENTION AND RETENTION SYSTEMS

Surface detention basins, either extended or dry ponds, are small impoundments usually serving drainage areas of 10 acres (4 ha) or less. Although a number of variations are possible, Figure 6-1 illustrates a typical detention basin profile. These systems are designed to store a portion of runoff and, through the use of restricting, gravity-flow outlet works, empty slowly following a storm event. In addition to attenuating peak runoff rates through redistribution of discharge hydrographs, detention of stormwater provides an opportunity for settling of suspended solids (e.g., sediment) and many other pollutants. Removal rates generally increase with detention time, defined by Haan et al. (1994) as the time difference between the centroids of the inflow and outflow hydrographs. Brown et al. (1996) report pollutant removal rates as high as 90 percent if stormwater is detained for 24 hours or more. Detention basins may also create benefits in terms of recreational use (e.g., golf course or parks) of seldom inundated portions of the pond and the promotion of wetland and wildlife habitat. Detention basins, however, require routine maintenance, including mowing, debris removal, and the eventual removal of sediment. Otherwise, they may become an aesthetic nuisance with respect to debris, odor, and insect-breeding.

Sizing a detention basin begins by estimating the volume of pond storage required. Storage criteria, along with the period over which the associated volume should be released, vary among storm and combined sewer authorities. Brown et al. (1996), however, recommend that, at a minimum, a basin store the equivalent of 0.5 in (13 mm) of runoff spread uniformly over the drainage area for water quality purposes. This volume is often referred to as the water quality volume (WQV). From a water quantity perspective, larger rainfall events (e.g., 5- or 10-year storms) should be used for design. The required volume is used to prepare a trial layout, including desired outlet configuration. Note that basic layout dimensions will often depend on site limiting factors such as topography, utilities, and geology of underlying soils. In addition, layout and outlet works may be adjusted so that the basin drains within a typical interstorm period of 72 hours, or 24 to 36 hours for landscaped basins. Once a trial design has been selected, storage-outflow and stage-outflow (i.e., water surface elevation-discharge) relationships are typically used to route an inflow hydrograph through the basin. Based on routing results, the size and detention time can be iteratively refined until a suitable design is attained. ASCE (1998) further recommends that this final volume be increased by 20 percent to allow for sediment accumulation in the pond.

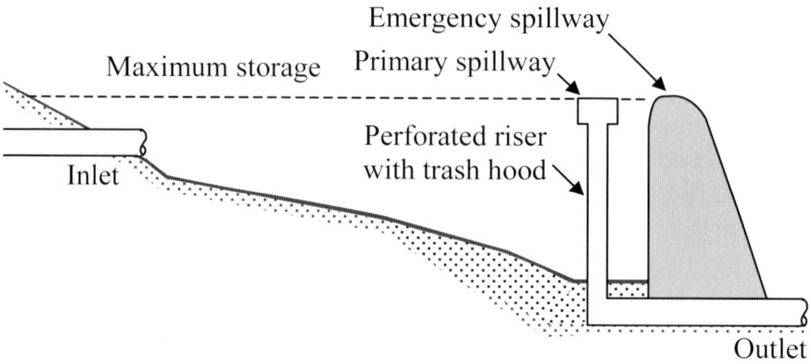

Figure 6-1: Typical detention basin

Table 6-1 lists a number of methods available for estimating a required detention storage volume, V_s. As an alternative, the modified rational method is commonly used for basins that are less than 30 acres (12 ha) in size (Chow et al., 1988). While there are several variations of the method, it generally assumes that the inflow hydrograph for the basin is trapezoidal in shape and can be constructed such that the period associated with the rising and recession limbs equals the time of concentration. Peak discharge, I_p, can be computed using the rational formula (see Chapter 2). The approach also assumes that the rising limb of the outflow hydrograph is linear and that its peak, O_p, falls on the recession limb of the inflow hydrograph, as shown in Figure 6-2. With similar assumptions, Aron and Kibler (1990) showed that the volume of required storage can be expressed as

$$V_s = \left[I_p t_r - O_p \left(\frac{t_r + t_c}{2} \right) \right] \times 60 \tag{6-1}$$

Table 6-1: Formulae for trial storage volumes

Relationship	Formula	Comments
Abt and Grigg (1978)	$\frac{V_s}{V_r} = \left(1 - \frac{O_p}{I_p}\right)^2$	Triangular inflow hydrograph and trapezoidal outflow hydrograph
Wycoff and Singh (1986)	$\frac{V_s}{V_r} = \frac{1.291\left(1 - \frac{O_p}{I_p}\right)^{0.753}}{\left(t_b/t_p\right)^{0.411}}$	Based on regression analysis
ASSHTO (1991)	$V_s = \frac{1}{2} t_b \left(I_p - O_p\right)$	Inflow and outflow hydrographs are triangular (i.e., $t_p = t_b/2$).
Kessler and Diskin (1991)	$\frac{V_s}{V_r} = 0.932 - 0.792 \frac{O_p}{I_p}$	For weir outlet and constant basin surface area; valid for $0.2 < \left(O_p/I_p\right) < 0.9$
Kessler and Diskin (1991)	$\frac{V_s}{V_r} = 0.872 - 0.861 \frac{O_p}{I_p}$	Orifice outlet and constant basin surface area; valid for $0.2 < \left(O_p/I_p\right) < 0.9$

Note: V_s is the required storage volume; I_p and O_p are the peak inflow and outflow (i.e., allowable outflow) rates of the basin, respectively; V_r is the total volume of runoff; t_b is the base time of the inflow hydrograph, defined for the Wycoff and Singh (1986) method as the time from the beginning of rise to a point on the recession limb where flow is five percent of I_p; and t_p refers to time to peak inflow.

where t_r and t_c are the design storm duration and time of concentration for the drainage area, both expressed in minutes. For the modified rational method, the design duration is that which maximizes basin storage volume and is frequently referred to as the critical duration. If using local intensity-duration-frequency (IDF) curves, its corresponding value can be obtained using an iterative technique. However, if IDF data is available in the form

$$i = \frac{a}{\left(b + t_r\right)^c} \tag{6-2}$$

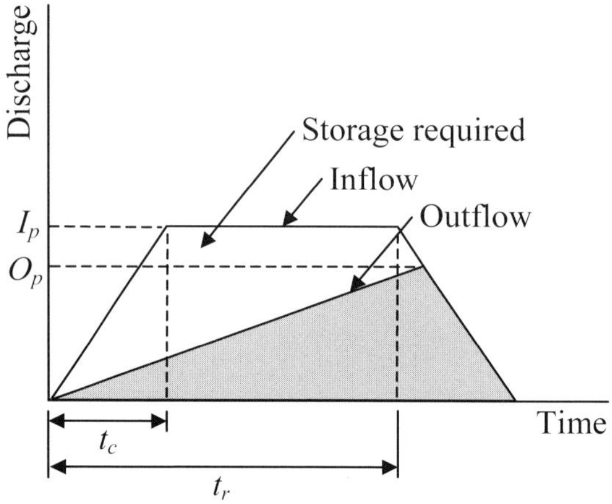

Figure 6-2: Inflow and outflow hydrographs for trial storage volume

where i is the average rainfall intensity and a, b, and c refer to curve fitting parameters, then the critical duration can be computed by differentiating Equation 6-1 and equating it to zero. Doing so yields the following expression:

$$\frac{a[b+t_r(1-c)]}{(b+t_r)^{c+1}}-\frac{O_p K_r}{2CA}=0 \qquad (6\text{-}3)$$

where, for time and intensity data given in minutes and in/hr or mm/hr, respectively, O_p is in cfs or m^3/s; C is a dimensionless runoff coefficient (see Chapter 2); A is the drainage area in ac or ha; and K_r is a conversion constant equal to 1.0 in U.S. customary units and 360 in S.I. units. Equation 6-3 can be solved for t_r, which is in turn used to evaluate the associated rainfall intensity and peak rate of flow entering the basin. Subsequently, Equation 6-1 can be used to evaluate a trial basin storage volume.

Another commonly used method for determining basin storage volume is that of the Federal Aviation Administration (FAA, 1970). Very similar to the modified rational method, the FAA approach incorporates a duration that produces the maximum basin storage volume. Alternatively expressed, the volume required is the maximum difference between cumulative inflow and outflow volumes to the basin, or

$$V_s = max\left(I_p t_r - mO_p t_r\right) \qquad (6\text{-}4)$$

where m is an outflow adjustment factor to account for the variation of outflow over time. It has been suggested that m should assume a value between 0.75 and 0.85 for most storms (Urbonas and Stahre, 1992). Guo (1999), however, showed that for hydrograph shapes similar to that defined in Figure 6-2,

$$m = \frac{1}{2}\left(1 + \frac{t_c}{t_r}\right) \tag{6-5}$$

This relationship is valid for durations that exceed the time of concentration and indicates that m ranges from 1.0 (i.e., when $t_r = t_c$) to 0.5.

Once a trial basin storage volume and configuration have been identified, a routing procedure should be used to evaluate the resulting outflow hydrograph. Two of the most common routing techniques are the level-pool method, also known as the Puls method, and the Runge-Kutta procedure. Compared to hydraulic methods that rely on St. Venant equations, these are both hydrologic, or lumped, approaches that are based on a simple mass balance of basin storage, S, or

$$\frac{dS}{dt} = I(t) - O(t) \tag{6-6}$$

where $I(t)$ and $O(t)$ represent time distributions of inflow and outflow (Mays, 2001). For level-pool routing, Equation 6-6 can be expressed in finite difference form as

$$\left(\frac{2S_{j+1}}{\Delta t} + O_{j+1}\right) = \left(I_j + I_{j+1}\right) + \left(\frac{2S_j}{\Delta t} - O_j\right) \tag{6-7}$$

where Δt is the time interval associated with the inflow hydrograph. The unknowns in Equation 6-7 are S_{j+1} and O_{j+1}, which can be determined by using a storage-outflow relationship, or more specifically, the relationship between $(2S/\Delta t + O)$ and O, which can be developed using stage-storage and stage-outflow functions. Stage-storage relationship can be derived from topographic maps or survey data, while a stage-outflow relation might be derived using hydraulic equations. For example, flow through weir-type outlets can be expressed as

$$O = CL\sqrt{2g}H^{3/2} \tag{6-8}$$

where C is a dimensionless weir coefficient; L is the effective length of the weir crest; g is the gravitational constant; and H is total head on the weir, or the sum of depth above the weir crest and velocity head of approach flow. Equation 6-8 applies to sharp crested weirs and uncontrolled ogee crests, as well as riser pipes, morning glory spillways, and gate controlled ogee crests when total head is low (i.e., weir flow). For the latter cases and for culvert outlets, stage, and thus total head, eventually reaches a condition such that orifice flow prevails, and discharge can be expressed as

$$O = C_o A_o \sqrt{2gH} \tag{6-9}$$

where C_o is a dimensionless orifice discharge coefficient; A_o is the orifice area; and H becomes the total head on the orifice. USBR (1987) provides additional guidance on these outlet relationships and on identifying stage-outflow relationships.

The Runge-Kutta method requires no storage-outflow function, but rather relationships between stage, h, basin surface area, A, and outflow. The scheme described here is third-order, whereby each hydrograph time interval is divided into three distinct periods and basin stage and outflow is computed for each period. This method is also more closely related to the hydraulics of flow through the basin (Chow et al., 1988). To begin, the change in storage in terms of stage can be expressed as

$$dS = A(h)dh \tag{6-10}$$

Substitution of this expression into Equation 6-6 and simplifying yields

$$\frac{dh}{dt} = \frac{I(t) - O(h)}{A(h)} \tag{6-11}$$

Equation 6-11 can be solved at each time step, j, using three approximations (i.e., Δh_1, Δh_2, and Δh_3) of Δh, where

$$h_{j+1} = h_j + \Delta h \tag{6-12}$$

and

$$\Delta h = \frac{\Delta h_1}{4} + \frac{3}{4}\Delta h_3 \tag{6-13}$$

The approximations are

$$\Delta h_1 = \left[\frac{I(t_j) - O(h_j)}{A(h_j)} \right] \Delta t \tag{6-14}$$

$$\Delta h_2 = \left[\frac{I\left(t_j + \frac{\Delta t}{3}\right) - O\left(h_j + \frac{\Delta h_1}{3}\right)}{A\left(h_j + \frac{\Delta h_1}{3}\right)} \right] \Delta t \tag{6-15}$$

and

$$\Delta h_3 = \left[\frac{I\left(t_j + \frac{2}{3}\Delta t\right) - O\left(h_j + \frac{2}{3}\Delta h_2\right)}{A\left(h_j + \frac{2}{3}\Delta h_2\right)} \right] \Delta t \tag{6-16}$$

The temporal variation of stage can in turn be used with the stage-outflow function to derive the outflow hydrograph, which can then be used to improve detention basin specifications (e.g., volume, dimensions, or outlet works).

Retention basins, or wet ponds, are very similar in concept to detention ponds, but they are generally larger in size and serve larger watersheds. They also release stored water at slower, more controlled rate than that of detention systems and maintain a permanent pool of water. In some cases, water may not be discharged at all, but instead be used as an emergency water supply and be consumed by plants, evaporate, or infiltrate into the underlying soil. Increased removal of stormwater pollutants (i.e., BOD, nutrients, sediment, and trace metals) is achieved through settling over longer detention times and by growth of phytoplankton, microorganisms, and shallow marsh plants. Generally, during wet weather, new runoff displaces existing water from the permanent pool, from which pollutants have already been removed. The new runoff is retained until it is displaced by subsequent loading. In this way, the permanent pool captures and treats the smaller and more frequently occurring runoff. Additional storage volume, above the permanent pool, is used to limit peak discharges caused by design storm events. Similar to detention facilities, criteria for determining the size of permanent pool vary among management authorities. Brown et al. (1996), however, recommend that the pool be at least three times the WQV specified for detention basins.

6.3 INFILTRATION BASINS

Infiltration basins are essentially depressed areas, either natural or excavated, into which stormwater is conveyed and allowed to percolate through the underlying permeable soil. In addition to reducing peak sewer loads, the infiltration process allows for removal of fine and soluble pollutants through adsorption, filtering, and microbial decomposition. Their appearance and construction is similar to that of detention ponds, although they can generally accommodate larger drainage areas of up to 50 ac (20 ha) (Brown et al., 1996). In fact, some facilities are designed as combined infiltration/detention basins. The major advantages of infiltration basins are that they replenish groundwater and locally preserve the natural water balance, and they are generally more cost effective than other BMPs. They do, however, experience a high failure rate due to unsuitability of soils and clogging, and they require frequent maintenance to prevent operational and nuisance problems.

Suitable infiltration basin operation requires that the facility be capable of capturing the design stormwater load and that the subsurface geometry and geology can sustain infiltration. With regard to the former, the basin can essentially be designed as detention pond, using previously described methods. Since water percolates slowly, however, a much larger surface area is typically required to disperse stormwater. Guo (2001) showed that the minimum basin surface area, A_b, can be estimated as

$$A_b \geq \frac{V_s}{Z(\phi - \theta_w)} \tag{6-17}$$

where ϕ is the soil porosity; θ_w is the specific retention, or initial water content, of the soil; Z is the distance between the basin bottom and the groundwater; and V_s is the required detention volume based on surface hydrology. This relationship is based on a conservative approach whereby it is assumed that the water storage volume available in the soil pores beneath the basin is greater than V_s. The maximum basin depth is equal to the denominator of Equation 6-17, and the associated drain time, T_d, of the basin is (Guo and Hughes, 2001)

$$T_d = \frac{Z(\phi - \theta_w)}{i_c} \tag{6-18}$$

where i_c is the final soil infiltration rate. As a general rule, for the basin to be feasible, soils lying at depths up to 5 ft (1.5 m) beneath the basin should have a minimum infiltration rate of 0.5 in/hr (13 mm/hr) (Brown et al., 1996); otherwise, the basin area required to limit peak runoff is excessive. In addition, the bottom of the basin should lie at least 5 to 10 ft (1.5 to 3 m) above the seasonally-high groundwater level and 5 ft (1.5 m) above bedrock (Guo, 2001).

6.4 SUBSURFACE DETENTION SYSTEMS

Particularly in areas where surface basins are not feasible due to space considerations, underground detention is an attractive alternative. Here, excess sewer loads are stored in large, buried tanks or conduits that are installed either in-line or off-line. In the former, the facility is integrated directly into the sewer network. For off-line systems, stormwater is collected before it enters the network and is discharged at a controlled rate to either a sewer or open watercourse. Design capacity is estimated using the same methods used for surface detention systems. Additional consideration should, however, be given to the structural design aspect of storage facilities.

6.5 SOLVED PROBLEMS

Problem 6.1 Detention volume (Abt and Grigg)

The peak runoff and time to peak discharge for a 20-ac drainage area are 120 cfs and 35 minutes, respectively. The excess rainfall from the design storm event is 2.8 in, spread evenly over the watershed. Use the Abt and Grigg (1978) method to estimate the detention basin storage required to limit the peak discharge to 35 cfs.

Solution

The volume of runoff is computed using the depth of excess rainfall,

$$V_r = \left(\frac{2.8}{12}\right)\left(20 \times 43{,}560\ ft^2/ac\right) = 203{,}280\ ft^3 = 4.67\ ac\text{-}ft$$

Using the Abt and Grigg (1978) formula, listed in Table 6-1,

$$\frac{V_s}{V_r} = \left(1 - \frac{35}{120}\right)^2 = 0.50$$

Then, the volume of storage required is

$$V_s = 0.50(4.67) = \underline{\underline{2.34\ ac\text{-}ft}}$$

Problem 6.2 Detention volume (Wycoff and Singh)

Solve Problem 6.1 using the Wycoff and Singh (1986) method.

Solution

The base time of the inflow is evaluated as the time at the beginning of runoff to a point on the recession limb where flow is five percent of I_p. Thus, for the triangular inflow hydrograph,

$$t_b = (2 \times 35) - 35\left[\frac{0.05(120)}{120}\right] = 68.25\ min.$$

Using the Wycoff and Singh (1986) formula,

$$\frac{V_s}{V_r} = \frac{1.291\left(1 - \frac{35}{120}\right)^{0.753}}{(68.25/35)^{0.411}} = 0.76$$

so that the volume of storage required is

$$V_s = 0.76(4.67) = \underline{\underline{3.55\ \ ac\text{-}ft}}$$

Problem 6.3 Detention volume (AASHTO)

Solve Problem 6.1 using the AASHTO (1991) method.

Solution

With the base time defined as twice the time to peak discharge, the AASHTO (1991) formula yields

$$V_s = \frac{1}{2}(70 \times 60\ sec/min)(120 - 35) = 178{,}500\ ft^3 = \underline{\underline{4.09\ ac\text{-}ft}}$$

Problem 6.4 Detention volume (Kessler and Diskin)

Assuming a weir outlet, solve Problem 6.1 using the Kessler and Diskin (1991) method.

Solution

From the information given in the problem statement, (O_p/I_p) = 0.3 and is within the acceptable range of 0.2 to 0.9. Then

$$\frac{V_s}{V_r} = 0.932 - 0.792\left(\frac{35}{120}\right) = 0.70$$

and the volume required is expressed as

$$V_s = 0.70(4.67) = \underline{\underline{3.27\ ac\text{-}ft}}$$

Problem 6.5 Detention volume (Kessler and Diskin)

Assuming an orifice outlet, solve Problem 6.1 using the Kessler and Diskin (1991) method.

Solution

Using the corresponding formula from Table 6-1 yields

$$\frac{V_s}{V_r} = 0.872 - 0.861\left(\frac{35}{120}\right) = 0.62$$

and the volume required is expressed as

$$V_s = 0.62(4.67) = \underline{\underline{2.90\ ac\text{-}ft}}$$

which is slightly lower than similar computations for a weir outlet.

Comments: Trial storage values range from 2.34 to 4.09 ac-ft, depending on the method of computation.

Problem 6.6 Detention volume (Modified rational)

Using the modified rational method, estimate the maximum detention basin storage required for a 30-ac drainage area having a runoff coefficient of 0.9 and a time of concentration of 25 minutes. The basin is intended to reduce the peak discharge to 25 cfs, and rainfall intensity (in/hr) for the area is expressed as

$\dfrac{101.2}{(15.6 + t_r)^{0.92}}$, where t_r is the rainfall duration in minutes.

Solution

Rainfall intensity is expressed in a form similar to that of Equation 6-2. Therefore, the fitting coefficients are a = 101.2, b = 15.6, and c = 0.92. Equation 6-3 can be expressed as

$$\frac{101.2[15.6 + t_r(0.08)]}{(15.6 + t_r)^{1.92}} - \frac{25}{2(0.9)(30)} = 0$$

Solving by trial-and-error, numerical analysis, or computational solver yields a critical storm duration of 64.71 minutes. The intensity associated with this duration is

$$i = \frac{101.2}{(15.6 + 64.71)^{0.92}} = 1.79 \ in/hr$$

and, using the rational equation, the peak inflow is

$$I_p = CiA = (0.9)(1.79)(30) = 48.3 \ cfs$$

The maximum volume of storage required can then be evaluated using Equation 6-1, or

$$V_s = \left[(48.3)(64.71) - 25\left(\frac{64.71 + 25}{2} \right) \right] \times 60 = 120{,}247 \ ft^3 = \underline{\underline{2.76 \ ac\text{-}ft}}$$

Problem 6.7 Detention volume (FAA)

Solve Problem 6.6 using the FAA (1970) method. Use the rational method to estimate peak runoff conveyed to the basin.

Solution

The FAA (1970) method used here integrates the modifications suggested by Guo (1999). The corresponding solution is based on Equation 6-4 and is summarized in Table P6-7.

Table P6-7

(1)	(2)	(3)	(4)	(5)	(6)
Duration (min)	Rainfall Intensity (in/hr)	Runoff Volume (ft^3)	*m*	Outflow Volume (ft^3)	Storage (ft^3)
40	2.51	162,648	0.81	48,600	114,048
45	2.32	169,128	0.78	52,650	116,478
50	2.16	174,960	0.75	56,250	118,710
55	2.02	179,982	0.73	60,225	119,757
60	1.89	183,708	0.71	63,900	119,808
65	1.78	187,434	0.69	67,275	120,159
68	1.72	189,475	0.68	69,360	120,115
69	1.71	191,144	0.68	70,380	**120,764**
70	1.69	191,646	0.68	71,400	120,246
75	1.60	194,400	0.67	75,375	119,025

For different assumed durations, the rainfall intensity is computed and used with the rational equation to evaluate runoff volume (i.e., $I_p t_r \times 60$) in Column 3. The outflow volume in Column 5 is computed as ($mO_p t_r \times 60$), noting that the value of m varies with duration as indicated by Equation 6-5. For example, for duration of 40 minutes, intensity and runoff volume are evaluated as

$$i = \frac{101.2}{(15.6+40)^{0.92}} = 2.51\ in/hr$$

$$I_p t_r = CiAt_r = (0.9)(2.51)(30)(40) \times 60\ \text{sec/min} = 162{,}648\ \text{ft}^3$$

The corresponding value of the adjustment factor, m, and the outflow volume are computed as

$$m = \frac{1}{2}\left(1 + \frac{25}{40}\right) = 0.81$$

$$mO_p t_r = 0.81(25)(40) \times 60\ sec/min = 48{,}600\ ft^3$$

The difference between inflow and outflow volumes, or 162,648 – 48,600 = 114,048 ft^3 in this case, is tabulated in Column 6. The maximum difference occurs at a duration of 69 minutes and is equal to 120,764 ft^3, or 2.77 ac-ft. This is approximately the same answer obtained for the previous problem, which is expected given the similarity of methods and the assumptions used.

Problem 6.8 Stage-discharge relationships

Consider a detention basin with two outlets. The lower outlet is an orifice having a diameter of 0.5 m and a discharge coefficient of 0.58. The upper outlet is a weir having a discharge coefficient of 0.37 and a crest length of 1 m. The centerline of the orifice lies at a stage of 60 m, while the weir crest is located at 62 m. Evaluate the stage-discharge relationship for this basin.

Solution

Column 1 of Table P6-8 lists the range of stages considered. Columns 2 and 4 indicate the corresponding head on the outlet devices. For example, at a stage of 62.2 m, the head on the weir is only 0.2 m since the crest lies at an elevation of 62 m, while the head on the orifice is 2.2 m. Using Equations 6.8 and 6.9, the discharge from each device is tabulated separately in Columns 3 and 4, the sum of which equals the total discharge in Column 5. Thus, the stage-discharge relationship is given in Columns 1 and 5.

Problem 6.9 Stage-discharge relationships

Under peak conditions, the stage in a newly-designed detention basin is expected to be at elevation 182 m. The basin outlet consists of a perforated riser pipe and is to discharges a mean flow of 0.003 m^3/s, averaged over the detention period. Assuming the 0.15-m diameter barrel of the outlet rests at an elevation of 180 m, estimate the number of 13-mm diameter perforations required.

Table P6-8

(1)	(2)	(3)	(4)	(5)	(6)
Stage, *h* (m)	Orifice		Weir		Total Discharge (m^3/s)
	Head, *H* (m)	Discharge (m^3/s)	Head, *H* (m)	Discharge (m^3/s)	
60.0	0.0	0.00	-	-	**0.00**
60.2	0.2	0.23	-	-	**0.23**
60.4	0.4	0.32	-	-	**0.32**
60.6	0.6	0.39	-	-	**0.39**
60.8	0.8	0.45	-	-	**0.45**
61.0	1.0	0.50	-	-	**0.50**
61.2	1.2	0.55	-	-	**0.55**
61.4	1.4	0.60	-	-	**0.60**
61.6	1.6	0.64	-	-	**0.64**
61.8	1.8	0.68	-	-	**0.68**
62.0	2.0	0.71	0.0	0.00	**0.71**
62.2	2.2	0.75	0.2	0.15	**0.89**
62.4	2.4	0.78	0.4	0.41	**1.20**
62.6	2.6	0.81	0.6	0.76	**1.58**
62.8	2.8	0.84	0.8	1.17	**2.02**
63.0	3.0	0.87	1.0	1.64	**2.51**
63.2	3.2	0.90	1.2	2.15	**3.06**
63.4	3.4	0.93	1.4	2.71	**3.64**
63.6	3.6	0.96	1.6	3.32	**4.27**
63.8	3.8	0.98	1.8	3.96	**4.94**
64.0	4.0	1.01	2.0	4.64	**5.64**

Solution

The average head on the barrel is approximated as

$$H = \frac{1}{2}(182 - 180 - 0.15) = 0.925\ m$$

Assuming each perforation behaves as an orifice and that a discharge coefficient of 0.4 applies, discharge could be expressed as

$$O = 0.003 = C_o A_o \sqrt{2gH} = (0.4)A_o\sqrt{2(9.81)(0.925)}$$

Solving for A_o yields a required area of 1.76×10^{-3} m^2. Then the number of perforations required is

$$\frac{1.76 \times 10^{-3}}{\pi(0.013)^2/4} = 13.3$$

To be conservative in detention time, 13 perforations, or holes, should be used, spaced evenly down the riser.

Problem 6.10 Level pool routing

The inflow hydrograph for a detention basin is triangular in shape, increasing linearly from zero to a peak of 360 cfs at 60 minutes and then returning to zero at 160 minutes. The basin storage increases linearly from zero at a stage, *h*, of 40 ft to 660,000 ft^2 at $h = 49$ ft, measured from mean sea level. The outlet is a low-level weir having its crest at an elevation of 40 ft and that can be described by the function $O = 13(h - 40)^{3/2}$, where *O* is expressed in cfs. If the pond is initially empty, determine the maximum discharge from the basin using the level pool routing technique.

Solution

As summarized in Table P6-10a, the relationship between storage and outflow for the basin must first be defined prior to routing computations. For different assumed values of stage, storage (Column 2) is first computed via linear interpolation of the given information. Outflow (Column 3) is computed as a function of stage using the provided weir relationship. The value of $\frac{2S}{\Delta t} + O$ is then computed in Column 4, using a time interval (i.e., Δt) of 10 minutes. Table P6-10b summarizes the stepwise evaluation of the new unit hydrograph (UH).

Table P6-10a

(1)	(2)	(3)	(4)
Stage, h (ft)	Storage (ft^3)	Discharge (cfs)	$\frac{2S}{\Delta t}+O$ (cfs)
40	0	0.0	0.0
41	73,333	13.0	257.4
42	146,667	36.8	525.7
43	220,000	67.5	800.8
44	293,333	104.0	1,081.8
45	366,667	145.3	1,367.5
46	440,000	191.1	1,657.8
47	513,333	240.8	1,951.9
48	586,667	294.2	2,249.8
49	660,000	351.0	2,551.0

Columns 1 and 2 list the inflow hydrograph. For the first time interval, S_1 and O_1 assume zero values, so that $\frac{2S_1}{\Delta t}-O_1$ is zero. The value of $\frac{2S_2}{\Delta t}+O_2$ (Column 5) at the beginning of the second time interval is given as

$$\left(\frac{2S_2}{\Delta t}+O_2\right)=(I_1+I_2)+\left(\frac{2S_1}{\Delta t}-O_1\right)=(0+60)+0=60\ cfs$$

This value is then used with linear interpolation of the values given in Table 6-10a to determine the outflow, O_2, which is listed in Column 6. In this case,

$$O_2=\left(\frac{13+0}{257.4-0}\right)\times 60=3\ cfs$$

The value in Column 4 for the next time step can be expressed as

$$\left(\frac{2S_2}{\Delta t}-O_2\right)=\left(\frac{2S_2}{\Delta t}+O_2\right)-2O_2=60-2(3)=54\ cfs$$

The computation then proceeds to the next time interval using a similar technique. From the values listed in Column 6, the maximum discharge from the basin is 293.7 cfs. Thus, the peak sewer loading rate was trimmed by approximately 20 percent from a value of 360 cfs.

Problem 6.11 Detention time

Using the definition provided by Haan et al. (1994), determine the detention time of the basin in Problem 6.10.

Solution

The centroid of the inflow and outflow hydrographs could be expressed as

$$\overline{t_I} = \frac{\sum_{j=1}^{J} t_j I_j}{\sum_{j=1}^{J} I_j}$$

where J refers to the total number of data points on the hydrograph. For the current problem, the denominator of this expression is equal to the sum of the entries in Column 2 of Table P6-10b, or 2,880 cfs. The numerator is

$$[(60)(10)+(120)(20)+(180)(30)+\ldots+(36)(150)]\times 60$$

which gives 12,672,000 ft^3. Then,

$$\frac{12{,}672{,}200}{2{,}880}\times\frac{1}{60}\ min/sec = 73.3\ min$$

Using a similar expression for the outflow hydrograph,

$$\overline{t_O} = \frac{\sum_{j=1}^{J} t_j O_j}{\sum_{j=1}^{J} O_j}$$

yields a centroid at 105.3 min. Thus, the detention time of the basin is 105.3 – 73.3, or 32 min. A detention time of this magnitude may suffice from a water quantity perspective, as shown in Problem 6.10, but would be inadequate from a water quality viewpoint. The drain time for the basin is equal to the time from when inflow first enters the basin to when outflow ceases. For this problem, if Table P6.10b were extended to when O_j returns to zero, drain time would be approximately 720 minutes, or 12 hours.

Table P6-10b

(1)	(2)	(3)	(4)	(5)	(6)
Time (min)	Inflow I_j (cfs)	$I_j + I_{j+1}$ (cfs)	$\frac{2S_j}{\Delta t} - O_j$ (cfs)	$\frac{2S_{j+1}}{\Delta t} + O_{j+1}$ (cfs)	Outflow O_j (ft^3)
0	0	-	-	-	0.0
10	60	60	0.0	60.0	3.0
20	120	180	54.0	234.0	11.8
30	180	300	210.4	510.4	45.3
40	240	420	458.8	878.8	109.1
50	300	540	790.0	1,330.0	167.9
60	360	660	1,195.6	1,855.6	234.4
70	324	684	1,668.2	2,352.2	287.5
80	288	612	2,114.6	2,726.6	**293.7**
90	252	540	2,451.2	2,991.2	285.7
100	216	468	2,689.0	3,157.0	250.4
110	180	396	2,838.2	3,234.2	233.9
120	144	324	2,907.6	3,231.6	193.5
130	108	252	2,905.2	3,157.2	174.0
140	72	180	2,838.2	3,018.2	134.5
150	36	108	2,713.4	2,821.4	113.6
160	0	36	2,536.4	2,572.4	76.3
170	0	0	2,312.6	2,312.6	59.2
180	0	0	2,079.0	2,079.0	36.6

Problem 6.12 Runge-Kutta routing

Consider a detention basin having a stage-surface area relationship as described below and discharge described by $O = 2.4h^{3/2}$. The basin is initially empty and is subjected to a triangular runoff hydrograph that reaches a peak of 7.8 m^3/s at 60 minutes and has a base time of 330 minutes. Use the Runge-Kutta method with a 30 minute time interval, Δt, to determine the maximum stage and outflow from the basin.

Stage, h (m)	0.0	0.5	1.0	1.5	2.0
Area, A (m^2)	10,000	14,906	19,812	24,718	29,624

Solution

The solution is summarized in Table P6-12, which shows the inflow hydrograph in Column 1 and the variation of stage and outflow in Columns 7 and 8. The time step is equal to 30 minutes, or 1,800 seconds. Since the basin is initially empty, $h_1 = O(h_1) = I(0) = 0$, and

$$\Delta h_1 = \left[\frac{0-0}{10{,}000}\right] \times 1{,}800 = 0$$

At 30 minutes, by interpolation, $I(0+30/3) = 1.3$ m^3/s, and since $\Delta h_1 = 0$,

$$\Delta h_2 = \left[\frac{1.3-0}{10{,}000}\right] \times 1{,}800 = 0.234$$

Then, $I(0+(2/3)\times 30) = 2.6$ m^3/s and by interpolation of the stage-area relationship and evaluation of the discharge relationship at $(h_1 + 2\Delta h_2/3)$,

$$\Delta h_3 = \left[\frac{2.6-0.148}{11{,}816.5}\right] \times 1{,}800 = 0.374$$

The value of Δh is evaluated from Equation 6-13 as

$$\Delta h_3 = \frac{0}{4} + \frac{3}{4}(0.374) = 0.281$$

and the corresponding stage at 30 minutes is 0 + 0.281 = 0.281 m. Substituting this value for h into the discharge expression gives

$$O = 2.4(0.281)^{3/2} = 0.357\,\frac{m^3}{s}$$

Computations then proceed to the next time interval in a similar manner. As shown in Table P6-12, the maximum stage and outflow are 1.562 m and 4.685 m^3/s, respectively.

Problem 6.13 Infiltration basin

A planned infiltration basin is to have a floor elevation of 102 ft and detain a volume, V_s, of 0.3 ac-ft. Field measurements indicate that the soil underlying the basin has a porosity of 0.39 and a specific retention of 0.10. If the groundwater table lies at elevation 91 ft, determine the minimum bottom area and maximum water depth for the basin.

Solution

Using the data provided, the minimum basin area is computed from Equation 6-17 as

$$A_b \geq \frac{0.3}{(102-91)(0.39-0.10)} = 0.094\ ac = \underline{\underline{4{,}097\ ft^2}}$$

which is approximately equal to a 64-ft square. The maximum pond depth is

$$(102-91)(0.39-0.10) = \underline{\underline{3.2\ ft}}$$

Table P6-12

(1)	(2)	(3)	(4)	(5)	(6)	(7)	(8)
Time (min)	Inflow $I(t_j)$ (m^3/s)	Δh_1 (m)	Δh_2 (m)	Δh_3 (m)	Δh (m)	h (m)	Outflow $O(t_j)$ (m^3/s)
0	0	-	-	-	-	0	0
30	3.90	0	0.234	0.374	0.281	0.281	0.357
60	7.80	0.483	0.544	0.563	0.543	0.824	1.795
90	6.93	0.587	0.454	0.377	0.430	1.254	3.370
120	6.07	0.296	0.232	0.187	0.214	1.468	4.269
150	5.20	0.141	0.101	0.071	0.089	1.557	4.663
180	4.33	0.041	0.014	-0.007	0.005	**1.562**	**4.685**
210	3.47	-0.027	-0.047	-0.061	-0.053	1.509	4.449
240	2.60	-0.076	-0.091	-0.102	-0.096	1.413	4.031
270	1.73	-0.114	-0.125	-0.135	-0.130	1.283	3.488
300	0.87	-0.145	-0.155	-0.163	-0.159	1.124	2.860
330	0	-0.173	-0.183	-0.192	-0.187	0.937	2.177
360	0	-0.203	-0.186	-0.173	-0.181	0.756	1.578
390	0	-0.159	-0.147	-0.136	-0.142	0.614	1.155
420	0	-0.126	-0.116	-0.107	-0.112	0.502	0.854
450	0	-0.099	-0.092	-0.085	-0.089	0.413	0.637
480	0	-0.079	-0.073	-0.067	-0.070	0.343	0.482

Problem 6.14 Infiltration basin

If the basin described in Problem 6.13 has an infiltration rate of 5.0 in/hr, determine the associated drain time of the basin.

Solution

From Equation 6-18,

$$T_d = \frac{(102-91)(0.39-0.10)}{5.0/12} = \underline{\underline{7.65\ hrs}}$$

This value is well within the typical interstorm period. Note that if clogging occurs, drain time increases accordingly. For example, if infiltration is reduced by 50 percent, drain time is doubled.

REFERENCES CITED

Abt, S.R. and N.S. Grigg, "An Approximate Method of Sizing Detention Reservoirs," *Water Res. Bull.*, no. 14, 956-965, 1978.

American Society of Civil Engineers (ASCE), *Urban Runoff Quality Management*, ASCE Manual of Practice No. 23, New York, NY,1998.

American Society of State Highway and Transportation Officials (AASHTO), *Model Drainage Manual*, Washington, DC, 1991.

Aron, G. and D.F. Kibler, "Pond Sizing for Rational Formula Hydrographs," *Water Res. Bull.*, vol. 26, no. 2, 255-258, 1990.

Brown, S.A., S.M. Stein, and J.C. Warner, *Urban Drainage Design Manual*, Hydraulic Engineering Circular No. 22, FHWA-SA-96-078, Federal Highway Administration, U.S. Department of Transportation, Washington, DC, 1996.

Chow, V.T., D.R. Maidment, and L.W. Mays, *Applied Hydrology*, McGraw-Hill, New York, NY, 1988.

Federal Aviation Administration (FAA), "Airport Drainage," Rep. 150/5320-5B, Washington, DC, 1970.

Guo, J.C.Y., "Detention Storage Volume for Small Urban Catchments," *J. of Water Res. Plng. and Mgmt.*, vol. 125, no. 6, 380-382, 1999.

Guo, J.C.Y. and W. Hughes, "Storage Volume and Overflow Risk for Infiltration Basin Design," *J. of Irr. and Drainage Engrg.*, vol. 127, no. 3, 170-175, 2001.

Guo, J.C.Y. "Design of Infiltration Basins for Stormwater," in *Stormwater Collection Systems Design Handbook*, ed. by L.W. Mays, McGraw Hill, New York, NY, 2001.

Haan, C.T., B.J. Barfield, and J.C. Hayes, *Design Hydrology and Sedimentology for Small Catchments*, Academic Press, New York, 1994.

Kessler, A. and M.H. Diskin, "The Efficiency Function of Detention Reservoirs in Urban Drainage Systems," *Water Res. Research*, vol. 27, no. 3, 253-258, 1991.

Mays, L.W., *Water Resources Engineering*, Wiley, New York, NY, 2001.

Urbonas, B. and P. Stahre, *Stormwater Best Management Practices and Detention*, Prentice-Hall, Englewood Cliffs, NJ, 1992.

U.S. Bureau of Reclamation (USBR), *Design of Small Dams*, 3rd ed., U.S. Department of the Interior, Denver, CO, 1987.

Wycoff, R.L. and U.P. Singh, "Preliminary Hydrologic Design of Small Detention Reservoirs," *Water Res. Bull.*, vol. 12, no. 2, 337-349, 1986.

CHAPTER SEVEN
Regulation of Sewer Systems

7.1 INTRODUCTION

When stormwater travels over impervious surfaces, the natural (i.e., pre-development) volume and rate of runoff to be collected are increased. In addition, the runoff is exposed to a host of pollutants, including oils, sediment, and lawn chemicals, that have accumulated on surfaces. These pollutants enter the collection system and, eventually, receiving waterways. Consequently, the U.S. Environmental Protection Agency (USEPA, 1999) has identified the following negative impacts caused by runoff:

- Altered hydraulic characteristics of receiving water bodies, including higher peak flows, increased frequency and duration of bankfull flows, increased occurrences of downstream flooding, and reduced base flow;
- Changes in receiving stream morphology, such as increased sediment transport capacity, leading to channel degradation and deposition further downstream, increased bank and shoreline erosion, and stream-widening;
- Public health and recreation impacts (e.g., increased risk of illness) due to contact with contaminated waterways, contamination of water supplies, beach closures, and fishing restrictions; and
- Negative impacts on aquatic habitat leading to changes in fish and macroinvertebrate populations and loss of species.

Similar and often more severe environmental and human health impacts are created by overflows from combined and sanitary sewers. During severe wet-weather events, the capacity of a combined sewer system or treatment plant may be exceeded, resulting in combined sewer overflows (CSOs). The magnitude of the overflow event volume is dictated by the volume and the duration of the runoff event. Similarly, sanitary sewer overflows (SSOs) are unintentional releases of sewage from sanitary systems; including overflows and basement backups. They are largely associated with sewer capacity exceedence by extraneous flows and infiltration/inflows (I/I), and are often caused by physical failure of the sewer, by improper system operation or management, or by extreme wet weather. Although SSOs occur mainly during wet weather as a result of I/I, they can also occur during dry weather. The solution to wet weather SSO problems often requires capital expenditures to reduce I/I, maximize and/or expand the capacity of conveyance and treatment systems. Dry weather SSOs are generally mitigated through operational improvements and increased maintenance. Ultimately, both CSOs and SSOs release varied mixtures of untreated wastewater, debris, and stormwater to nearby watercourses or onto streets and other public areas.

The U.S. federal government has created a number of regulatory programs aimed at reducing the negative impacts associated with urban stormwater runoff and the occurrences of CSOs and SSOs. These programs establish a consistent basis and guidance for the design, operation, and management of sewer collection systems and related runoff controls.

7.2 LEGAL AUTHORITY

The U.S. Congress passed the Clean Water Act (CWA) (Public Law 92-500) in 1972 to protect the nation's water resources by regulating discharges to waterways. This legislation has guided water pollution control efforts since its inception and continues to be the basis for additional programs and policies aimed at managing polluted stormwater runoff, reducing CSOs and SSOs, and preserving or restoring receiving waters. The CWA established the National Pollutant Discharge Elimination System (NPDES), which requires that any entity wishing to discharge pollutants from a point source (i.e., discrete, identifiable source such as a conduit, channel, or container) must obtain a permit to do so. The permit limits the amount of discharge and the pollutant concentration it contains.

7.3 NATIONAL STORMWATER PROGRAM

Along with cooperation from state and local governments, the USEPA has developed and currently directs the National Stormwater Program (NSP). The program is part of the Water Quality Act of 1987, an amendment of the CWA that calls for a two-phased approach to the implementation of measures that reduce stormwater impacts.

7.3.1 Phase I

The first phase of the NSP was implemented in 1992 and relies heavily on the NPDES permit program to regulate discharges from those sources having the greatest potential to negatively impact receiving waters. Permits are required for certain industrial activities, for construction sites that disturb five acres (2 ha) or more of soil, and for discharges from medium- to large-sized (i.e., serving an incorporated community or county having a population of 100,000 or more) municipal separate storm sewer systems (MS4s) (USEPA, 2000a).

When approved by the USEPA, the NPDES program can be administered by corresponding state governments. Approval typically depends upon the ability of the state to show that sufficient resources and capabilities exist to effectively manage and enforce permits. Local governments also play an important role in Phase I policy. Municipalities are responsible for obtaining permits, for maintaining operations for their own collection systems, and for recordkeeping, inspection, and enforcement of permits for certain construction and industrial activities under their authority (Dodson and Maske, 2001).

Note that agricultural runoff and most other non-point source activities are generally exempt from NPDES requirements. Runoff from these sources, however, is covered under section 303(d) of the CWA. This provision requires each state to develop a list of lakes and streams that are impaired or threatened according to state standards and designated uses. Listed waters are then targeted for pollution reduction through development of Total Maximum Daily Loads (TMDLs), which set limits on pollutant loads from contributing point discharges and non-point sources.

7.3.2 Phase II

In 2003, Phase II of the NSP essentially expanded the permitting policy to construction sites that disturb one acre (0.4 ha) or more of soil, to all urban stormwater collection systems, and to sewer systems outside of urban areas that serve populations of 10,000 or more and have a population density of at least 1,000 persons/mi^2 (385 persons/km^2). Specifically, Phase II requires MS4s to develop, implement, and enforce best management practices (BMPs) (see Chapter 8) and measurable goals for the following six minimum controls (USEPA, 2000b):

- A public education program must be undertaken to distribute educational materials or conduct outreach activities concerning the impacts of stormwater. The program should emphasize roles the public can assume to reduce these impacts, such as ensuring proper use of lawn and garden chemicals, protection of riparian vegetation, and proper disposal of household hazardous wastes. It is recommended that the educational activities target specific public communities, as well as commercial, industrial, and institutional entities.

- MS4s must develop, implement, and enforce a program to detect and eliminate illicit discharges, effectively defined as any discharge not covered by a current NPDES permit. This program should involve mapping of the corresponding system and waters, formally prohibiting non-stormwater discharges to sewer systems, and addressing illicit discharges by enacting a detection system and by informing the public of the hazards associated with improper waste disposal.

- A plan must be developed, implemented, and enforced to reduce pollutants from construction activities. The plan must include a regulatory mechanism that requires site operators to use erosion and sediment controls, along with other BMPs, and to control the improper disposal of construction-related waste. It must also include procedures for site plan review, for consideration of information submitted by the public, and for site inspection and enforcement of control measures.
- A program must be established to address concerns related to runoff generated from new developments. The program should rely upon a combination of BMPs, adopt an ordinance to address post-construction runoff, and ensure adequate long-term operation and maintenance of BMPs.
- MS4s must develop and implement an operation and maintenance plan that includes personnel training and ultimately prevents or reduces runoff from municipal activities.
- The public should be involved in the development, implementation, and review of a stormwater management program. The participation process should make efforts to reach all economic and ethnic groups.

Additional requirements, such as spill prevention and response, monitoring of wet and dry weather flows, and special inspection and enforcement of high-risk activities, may be imposed for larger stormwater collection systems. When implemented together, these controls are designed to reduce, to the maximum extent possible, the quantity of pollutants discharged into waterways and limit the associated hydraulic impact. It is important to recognize that selection of the best combination of BMPs for a particular site is a subjective matter that depends on pollutants, climate, condition of receiving waters, local concerns, physical features, and other components of a comprehensive watershed-based plan.

7.4 COMBINED SEWER OVERFLOW CONTROL POLICY

The USEPA's CSO Control Policy is designed to guide permitting authorities and municipalities that own or operate combined sewer systems in meeting pollutant reduction goals. Published in 1994, the policy operates fully through the NPDES permitting program. Thus, it does not alone establish or affect legal rights or obligations; the ultimate obligation of permittees is to meet the requirements of the CWA (Moffa, 1997). The policy emphasizes the following four basic principles in controlling CSOs:

- Clear levels of control to meet health and environmental objectives;
- Flexibility to consider the site-specific nature of CSOs and determine the most cost-effective method for their control;
- Phased implementation of CSO controls to accommodate the financial capacity of communities; and
- Revision of water quality standards during the development of plans to reflect the site-specific impacts of CSOs.

The policy sets forth clear expectations for permitting authorities and municipalities. These include the implementation of at least nine technology-based controls and the development of a long-term control plan (LTCP) for implementation and monitoring of the controls, ultimately leading to full compliance with the CWA.

7.4.1 Nine Minimum Controls

The first major deadline under the CSO Control Policy was in 1997, when the Nine Minimum Controls (NMCs) had to be implemented. These controls are meant to be positive measures that communities can implement in a relatively short period without requiring significant engineering investigation or construction. As with stormwater programs, the number and combination of BMPs and other measures to be implemented in satisfying the minimum controls vary widely and depend on site-specific issues. The USEPA-prescribed list of NMCs include (USEPA, 1995):

- Establishing a proper operation, maintenance, and inspection program for the sewer collection system to ensure that it functions in a way that maximizes treatment of combined waste and still complies with NPDES permit restrictions. By enabling existing facilities to operate as effectively as possible, the magnitude, frequency, and duration of CSOs should be reduced.
- Making modifications to the sewer system to maximize use of the collection system for storage, essentially enabling the system to store wet-weather flows until downstream sewers and facilities can accommodate them. Modifications can be simple in nature, including, but not limited to, removal of flow obstructions (e.g., sediment and debris) that limit storage capability, correcting deficiencies created by improper maintenance, tide gate maintenance and repair, adjustment of

regulator settings, retardance of inflows by using special gratings, or use of localized upstream detention.

- Determining whether non-domestic sources (i.e., industrial and commercial sources) are contributing to CSO impacts and, if so, investigating ways to control them. The objective of this control is to minimize the negative impacts and occurrences of CSOs by modifying inspection, reporting, and oversight procedures within the approved pretreatment program.
- Examining and modifying the sewer system and treatment plant if needed to enable as much wet-weather flow as possible to reach the treatment facility. This control will typically require an evaluation of the ability of the facility to operate acceptably at incremental increases in wet-weather flows.
- Eliminating CSOs during dry weather by improving operation and maintenance programs or by making physical changes to regulator and overflow devices. Since the NPDES program prohibits dry weather overflows, this control is directly enforceable under the CWA.
- Controlling, if not eliminating, visible solids and floatable material in CSOs by using measures such as source pollution reduction or by installing devices such as baffles, screens, and skimmers. Note that several other minimum controls (i.e., increased storage of collection systems and maximization of flow to treatment facilities) are likely to assist in the reduction of solids and floatable matter as well.
- Implementing a pollution prevention plan intended to reduce, to the greatest extent possible, the amount of pollutants that enter collection systems. The program should include BMPs such as street sweeping, public education programs, solid and hazardous waste collection, and recycling.
- Informing the public of the location and occurrences of CSOs, their possible health and environmental impacts, and the recreational or commercial activities (e.g., swimming and shellfish harvesting) that have been curtailed as a result of CSOs. Acceptable methods for notification include postings at affected areas or select public places, media notices, telephone hotlines, and letters to affected residents.
- Monitoring, using visual inspections, automated measurement, or other methods to effectively characterize and document CSO impacts and the efficacy of CSO controls. This control is a precursor to the more extensive characterization and monitoring efforts that need to be conducted as part of the LTCP.

7.5 SANITARY SEWER OVERFLOW CONTROL POLICY

Historically, municipalities with a separate sewer system that carries only wastewater have not been required to obtain an NPDES permit. The justification for this is that permits are designed only for entities that discharge pollutants from point sources. Since sanitary sewers do not have any discharge pipes or designated outfalls to waterways, any and all SSOs are considered illegal under the CWA. However, the USEPA acknowledges that occasional overflows occur in many systems as a result of wet-weather infiltration and inflow, undersized, malfunctioning, or failing systems, improper management, backups in sewer service connections, or a combination of these factors.

In response, the USEPA has proposed a SSO Control Policy which would require corresponding communities to obtain NPDES permits for their sanitary sewer systems (USEPA, 2001). The draft policy was initially published in the Federal Register in 1999, but was subsequently withdrawn for further review. If adopted, the goal of this policy would be to improve sewer system operation, reduce the frequency and occurrence of SSOs, and to better inform the public about SSOs and their impacts. The final SSO Control Policy will likely include the following four major elements:

- All communities with satellite collection systems, which are those systems whose owner or operator is not the owner or operator of the treatment facility, would be required to obtain and comply with an NPDES permit;
- As part of the NPDES permit, municipalities would be required to implement a Capacity Assurance, Management, Operation, and Maintenance (CMOM) program;
- Although the CWA prohibits all SSOs, communities would have limited protection (i.e., affirmative defense) from enforcement in cases when SSOs are caused by severe and unusual weather events (e.g., hurricanes) that are beyond reasonable control of the operating entity; and
- Municipalities must develop record-keeping, reporting, and public notification programs that, in particular, inform the public regarding local occurrences and frequency of SSOs.

In anticipation that the proposed policy will be approved, the USEPA has increased enforcement activities related to sanitary sewer systems and many permit granting authorities are already writing portions of the anticipated policy into NPDES permits.

7.5.1 CMOM Program

The primary component of the proposed SSO Control Policy is the CMOM provision, which serves as the basis for developing permit requirements for improvement projects and operation and maintenance programs. The general requirements of CMOM are that permit holders must develop a customized plan to prevent and mitigate SSOs by

- Effectively operating and maintaining the sewer system;
- Providing sufficient collection system capacity;
- Responding promptly and effectively to mitigate overflow events;
- Notifying parties affected by overflows; and
- Making the CMOM plan available to the public.

The plan should begin with a self-audit to determine the locations and causes of SSOs during normal system operation. The plan must then delineate organizational structure and assignment of responsibilities, list objectives, goals, and legal authority, and clearly indicate the procedure for reporting SSO events. It must also address maintenance programs and practices, parts inventory and availability, engineering documentation of the system itself, including standards for design, construction, and inspection, and personnel training. It is expected that the SSO Policy will require that the CMOM plan be monitored and updated periodically.

Two subordinate parts of the CMOM plan must also be prepared. The first is an Overflow Response Plan, which lists the operating procedures that ensure rapid detection and response to SSOs and communicating corresponding information with all affected parties. The second, which can be a much more substantial undertaking, is a System Evaluation and Capacity Plan. This plan will require the collection and management of information and data about the sewer and should discuss the capital improvements or rehabilitation efforts that are necessary to bring the sewer system into compliance. In many cases, extensive computer simulation will be required to assess capacity and predict potential overflow locations.

As part of a cooperative agreement with the USEPA, the American Society of Civil Engineers has developed a guidance document that contains more than 100 measures, or solutions, that can be implemented to prevent and mitigate SSOs (ASCE, 2004). Solutions are organized into categories that correspond to the CMOM program: capacity solutions, management solutions, and operation and maintenance solutions.

Capacity solutions can be described as capacity assurance planning, including modeling, master planning, or capacity enhancement through system modification or capital projects. These projects may include increasing conveyance capacity or the storing of peak flows until the existing system can safely convey wet-weather flow. Management solutions involve improvements in the managing of assets, organizational structure, information, resources, and finances. In addition, these solutions could include benchmarking, revision of legal ordinances, updating standards, permitting, progress monitoring and periodic self-assessment, development of overflow response plans, and training and certification of personnel. Operation and maintenance solutions range from proactive to corrective measures, including hydraulic cleaning and root removal. While the nature of solutions may vary, nearly all will require good records, effective communication, thoughtful planning, and sound measurement tools to be effective. Furthermore, it is the intent of the USEPA that different solutions from the three categories be innovatively combined to derive a comprehensive approach to SSO reduction.

REFERENCES CITED

American Society of Civil Engineers (ASCE), *Sanitary Sewer Overflow Solutions: Guidance Manual*, ASCE, Reston, VA, 2004.

Dodson, R.D., and C.T. Maske, "Regulation of Stormwater Collection Systems in the United States," in *Stormwater Collection Systems Design Handbook*, ed. by L.W. Mays, McGraw Hill, New York, NY, 2001.

Moffa, P.E., "The Combined Sewer Overflow Problem: An Overview," in *Control and Treatment of Combined Sewer Overflows*, ed. by P.E. Moffa, Van Nostrand Reinhold, New York, NY, 1997.

U.S. Environmental Protection Agency (USEPA), *Combined Sewer Overflows: Guidance for Nine Minimum Controls*, EPA 832-B-95-003, Washington, DC, 1995.

U.S. Environmental Protection Agency (USEPA), *Preliminary Data Summary of Urban Storm Water Best Management Practices*, EPA 821-R-99-012, Washington, DC, 1999.

U.S. Environmental Protection Agency (USEPA), *Storm Water Phase II: Final Rule – An Overview*, EPA 833-F-00-001, Washington, DC, 2000a.

U.S. Environmental Protection Agency (USEPA), *Storm Water Phase II: Compliance Assistance Guide*, EPA 833-R-00-002, Washington, DC, 2000b.

U.S. Environmental Protection Agency (USEPA), *Proposed Rule to Protect Communities from Overflowing Sewers*, EPA 833-F-01-001, Washington, DC, 2001.

CHAPTER EIGHT
Management of Urban Watersheds

8.1 INTRODUCTION

Preventing and controlling the problems associated with stormwater runoff and wet-weather sewer overflows are challenging tasks. Many of the respective impacts are transient and elusive, especially in the case of pollutant discharges, and sources of impacts are not easily identified. Furthermore, impacts are often interrelated and cumulative. For example, both degraded water quality and increased peak discharges combine to negatively affect a habitat and biological resources.

Mitigation of such complex problems begins with a carefully prepared, yet evolving, watershed management plan. The plan must investigate and take inventory of all potential environmental and health impacts and subsequently identify best management practices (BMPs) required to meet specified water quality and quantity goals. Unfortunately, just as there is rarely a clear answer as to which combination of BMPs will yield the best result, there is no single approach to watershed management. More typically, there are a number of different programs and possible methods that are all based around a central principle that protection of water quality and quantity can best be achieved if all stakeholders (i.e., communities, private organizations, and regulatory agencies) work together.

Although the discussion that follows identifies some key elements of urban watershed management, it is not meant to be an all inclusive guide for managers and engineers. Over the last decade, a prolific amount of literature has surfaced regarding the various aspects of effective watershed management; the information ranges from technical design guidelines for control measures to more qualitative analyses of related social, political, and economic issues. A number of case studies have also been published, some of which detail success stories and others that describe failed attempts at watershed management. Additionally, the reader should identify with the underlying message conveyed through this Chapter; that water resources professionals should develop an understanding of the importance of comprehensive watershed planning and management and learn from the past to more effectively control the negative impacts associated with stormwater.

8.2 BENEFITS OF A WATERSHED APPROACH

Historically, the approach to water quality protection and the maintenance of natural runoff rates has been a patchwork of legislation and control measures that focus on individual sources of problems. A watershed approach, however, can more effectively resolve stormwater and other wet-weather problems. The approach is very similar in concept to what Pearse et al. (1985) have referred to as "comprehensive watershed management," which involves:

- A plan for the future of the watershed that considers and integrates all uses of a basin's water resources and other activities that affect water quantity and quantity;
- Information about a watershed's hydrologic regime;
- Modeling that is capable of revealing the full range of impacts that would be produced by particular uses and developments in the watershed;
- Specified management objectives for the entire watershed and criteria for assessing management alternatives in an objective and unbiased way;
- Participation of all relevant regulatory agencies; and
- Provisions for public participation in determining multiple objectives and evaluating management decisions.

This type of holistic approach bases decisions on a spatial scale that corresponds to a drainage basin. Thus, it provides a more logical and more effective basis for water resource management decisions than political boundaries. It also ensures that affected communities, regulatory agencies, and other local stakeholders have an understanding of their individual and collective roles in mitigating impacts and allows concerns to be addressed synergistically with greater economy and efficiency. Since the early 1990s, the U.S. Environmental Protection Agency (USEPA) has publicly recognized the importance of such watershed-scale planning as part of water quantity and quality control efforts. Unfortunately, in most cases, describing the approach is far easier than its successful implementation. Conflicting stakeholder objectives, entrenched opinions and past relationships, funding difficulties, and numerous uncertainties are only a few of the constraints that must be overcome if benefits are to be realized.

8.3 DEFINING THE PROBLEM AND OBJECTIVES

General impairments caused by urban runoff and sewer overflows range from reduced water quality due to the accumulation and transport of particular pollutants to stream channel erosion and localized flooding caused by increased volumes and rates of runoff. Identifying and ranking specific problems associated with these concerns for a given watershed begins by conducting a basin-wide inventory of natural resources (e.g., waterways, wetlands, fish and wildlife, endangered species, and cultural resources). For each resource, possible users (e.g., commercial navigation, agriculture, potable water supply) should be designated, existing conditions should be quantified, and, to the extent possible, current and future potential sources of pollution should be identified. This initial step will simultaneously serve as a benchmark to measure the effectiveness of future control measures that are implemented. Modeling studies will likely be needed at this stage in order to better understand hydrologic and hydraulic behavior of the watershed in response to both large and smaller, more frequent storm events. Finally, it is important to identify those individuals or groups (i.e., stakeholders) that are in some way affected by or play a role in regulating the problems identified.

Similar to the fact that the problem can take many forms, the goals of urban watershed management can vary widely. Goals can generally be based on multiple criteria or can be focused on singular objectives (e.g., reducing peak flows at one or more locations or improving specific pollutant removal efficiency to meet Total Maximum Daily Load criteria). Moreover, goals will likely differ according to stakeholder. It is important for these stakeholders to convene and collectively evaluate what is envisioned for the future of the watershed, how that vision can be realized, and who will fund corresponding efforts.

8.4 BEST MANAGEMENT PRACTICES

BMPs represent certain abatement procedures, devices, restrictions, or management activities that can be implemented to more effectively achieve watershed management goals. While some BMPs focus on either peak flow reduction or pollution control, many can be considered dual-purpose in that they provide both water quality and quantity benefits. In addition, specific BMPs can be designed and implemented separately or can be applied in a locally-strategic combination (i.e., treatment train) to more cost-effectively meet related goals.

Numerous BMPs for urban runoff water quality and quantity control have been, and continue to be, developed. In order to enable the transfer of evolving technology to practitioners, the USEPA and the American Society of Civil

Engineers initiated the National Stormwater BMP Database project in the late 1990s (ASCE, 2004). The initial database documented over 800 BMPs, and data for more BMPs are being added regularly as part of this ongoing project. Representative information for the BMPs includes test site location, watershed behavior, climate statistics, design parameters, and recommended monitoring data and equipment.

8.4.1 Types of BMPs

BMPs can generally be categorized as either source control and prevention techniques or passive treatment measures. For a variety of reasons (e.g., cost and site suitability), some of these practices are much more widely used than others. This discussion describes some of the more common BMPs and highlights their advantages and disadvantages.

8.4.1.1 Pollution Control and Prevention

Once polluted runoff enters collection systems or receiving waters, it is more difficult and costly to mitigate its impact. As a result, the USEPA has placed a priority on implementing control measures that reduce possible pollutant discharges or peak flows at the source. Best results are likely to be achieved when such controls are implemented on a watershed-scale, rather than on a site-specific or individual basis. Some of the more common examples of pollution control and prevention BMPs include public education, land use planning, improved landscaping, and maintenance techniques.

Public Education and Outreach: The public is often unaware of the negative effects of their actions with respect to pollution. Proper education on day-to-day activities that can be undertaken (e.g., recycling of automotive fluids, household chemical and fertilizer use and disposal, and animal waste control) is an effective method for limiting the amount of pollutants that enter collection systems and streams (USEPA, 1999a). Education can be formalized (i.e., workshops) or can be achieved through the distribution of pamphlets and other types of public service announcements.

Land Use Planning and Conservation: Prior to urban development, thoughtful planning that considers multiple objectives, including environmental impacts, is an important preemptive step in controlling problems associated with runoff. Low-impact development strategies may involve minimization of directly-connected impervious surfaces (i.e., those draining directly to a collection system), application of site depressions and rain gardens, implementing tax incentives and zoning ordinances for directed growth, and formalizing the protection of sensitive areas. In most cases, high-density, clustered urban development is preferred over low-density, suburban development as it tends to limit the creation of extensive impervious cover on a regional scale, thus minimizing impacts to numerous subwatersheds.

Landscaping and Vegetative Cover: Landscape planning should avoid long, steep slopes and provide terracing, contouring, and drainage channels as needed to limit runoff and erosion. Bare areas should be covered with vegetation or material such as hay or mulch. Doing so not only reduces erosion potential, but the increased surface roughness will improve infiltration to underlying soil and reduce runoff velocity. In places where bare areas are temporarily unavoidable (i.e., construction activities), straw bales, check dams, and silt fences are often used as sediment traps to prevent eroded material from entering nearby collection systems or streams. In addition, cities or drainage districts often obtain drainage easements for smaller channels to be modified as part of urbanization efforts and commonly install vegetative liners or riprap within channels. According to the Federal Highway Administration (Normann, 1975), the median diameter, D_{50}, of riprap on the bed of a sloping channel should be approximately

$$D_{50} = \frac{\gamma dS}{5} \tag{8-1}$$

where γ is the specific weight of water in lbs/ft^3; d is the maximum stable depth of flow in ft; and S is the slope of channel. Riprap should then be graded in size from $0.2 \times D_{50}$ to $2.0 \times D_{50}$ (Wurbs and James, 2002).

Maintenance and Housekeeping: After urbanization has occurred, a number of ongoing practices are important to prevent pollutants from entering collection systems and receiving waters. These practices include, among others, street sweeping, implementing formal waste collection programs, removal of trash and other debris from catch basins, maintenance and stabilization of roadways and ditches, and use of improved roadway deicing methods.

8.4.1.2 Stormwater Treatment

Stormwater treatment controls, sometimes referred to as structural BMPs, are those that generally do not require active operational control, but only routine maintenance. These BMPs rely upon key mechanisms of settling, filtration or infiltration, sorption, biodegradation, and/or evapotranspiration. Examples include various stormwater basins (e.g., detention, retention, and infiltration), filter strips, sand filters, and swales. Modeling of these systems typically relies on first principles of hydrology and environmental engineering (i.e., unit processes); it is important to note, however, that there still exists significant room for improving the prediction accuracy of current methods and models.

Detention and Retention Basins: These ponds are the most common passive treatment processes due to their versatility in treating a variety of pollutants and capacity to reduce downstream flooding. Detention facilities typically focus on attenuation of peak flows, while retention basins can serve both water quantity and quality purposes. A number of different designs exist, but for most applications, their main disadvantage is that they require roughly 10 to 20 percent of the basin's area (ASCE, 2001). Generally, only moderate to low maintenance (e.g., sediment removal) is required to ensure proper operation of basins. For further information regarding the design of these BMPs, refer to the information presented in Chapter 6.

Filter (Buffer) Strips: As shown in Figure 8-1, zones of dense vegetation ranging from grass to forest can be designed and planted on contoured areas or in riparian areas, perpendicular to overland flow, to attenuate flows and filter particulates from runoff. Pollutant removal in these systems occurs through detention, filtration by vegetation, and infiltration into underlying soils. Filter strips are generally limited to drainage areas of 5 ac (2 ha) or less and are often used as pretreatment for other BMPs (Brown et al., 1996). Special care should be taken so that flow does not prematurely form a concentrated channel and thus short-circuit the filter.

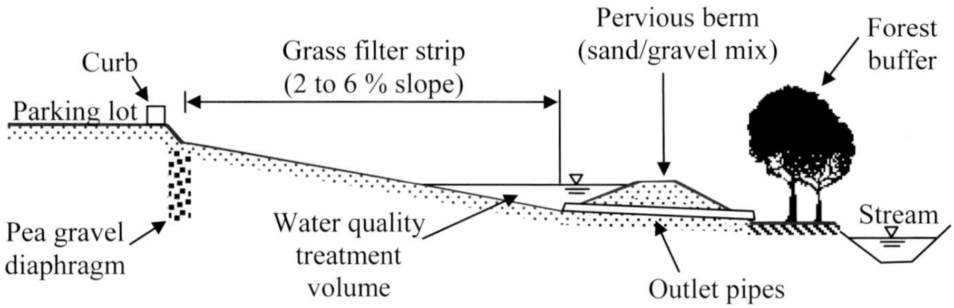

Figure 8-1: Typical filter strip (adapted from Claytor and Schueler, 1996)

As a guide, strips should have hydraulic retention times of five minutes or more; should not incur flow velocities greater than 0.9 fps (0.3 m/s); have a Manning roughness coefficient, n, in the range of 0.2 to 0.24; and should maintain flow depths that are less than 1.0 inch (25 mm) (ASCE, 1998). These criteria can be used to determine the minimum filter dimensions. First, the maximum unit-width loading, q, is determined using the Manning equation for overland flow, expressed as

$$q = \frac{K_m}{n} h^{5/3} S^{1/2} \tag{8-2}$$

where q is in cfs (m^3/s); K_m is a constant equal to 1.49 in U.S. customary units and 1.0 in S.I. units; n is the Manning roughness coefficient (see Chapter 2); h is the allowable depth in ft (m); and S is the dimensionless slope of the filter strip. The minimum width can be then computed by dividing the design flow rate by q, and the corresponding required length can be explicitly estimated by rearranging the kinematic wave equation for overland flow as follows:

$$L_f = \frac{t_t^{5/3} i^{2/3} S^{1/2}}{K_f n} \tag{8-3}$$

where L_f is in ft (m); t_t is the time of concentration through the filter, assumed to be equivalent to the retention time, in minutes; i is the rainfall intensity in in/hr (mm/hr); and K_f is a constant equal to 0.899 in U.S. customary units and 25.55 in S.I. units. Since Equation 8-3 assumes a constant Manning n and a homogeneous catchment and rainfall intensity, results are approximate. Other time of concentration relationships for overland flow could be used to estimate length as well (see Chapter 2); again, however, attention should be paid to assumptions used in their development and to their limitations. Note that regardless of the value computed using Equation 8-3, lengths should be no less than 20 ft (6 m) (ASCE, 1998). Furthermore, if runoff is heavily polluted and the filter is in a public area (e.g., park or playground), steps should be taken so that the contaminated sediments trapped in the filter do not pose a human exposure hazard. Generally speaking, filter strips have lower maintenance costs than other BMPs, especially if the natural landscape and vegetation is used in design. Unfortunately, they require significant land area, sometimes consuming more than 25 percent of the corresponding drainage area.

Sand Filters: Although variations exist, these systems generally utilize a bed of sand or similar media to filter sediment and pollutants from runoff prior to infiltration to native soils or being returned to a stream through underdrains (Brown et al., 1996). The overall design typically includes a pretreatment (i.e., sedimentation) basin and flow spreader to capture runoff and distribute it slowly to the filter. Filters are generally very adaptable – they can be used in areas with thin soils, high evaporation rates, low infiltration rates, and limited space – and can serve watersheds up to 30 ac (12 ha) in size, depending on the design (ASCE, 2001). In application, the top of the filter media must be completely horizontal to prevent disproportionate use of the media, and the sand bed should be at least 1.5 ft (0.5 m) deep. The minimum required surface area of the bed, A_b, can be computed as (Debo and Reese, 1995)

$$A_b = \frac{(V_c)Z_b}{i_c(h_{max} + Z_b)D_T} \tag{8-4}$$

where V_c is the capture volume, taken to be the water quality volume, or first flush, which is the equivalent of 0.5 in (13 mm) of runoff spread uniformly over the impervious drainage area; Z_b is the depth of the sand bed; K_p is the infiltration capacity; h_{max} is the maximum depth of water over the filter surface; and D_T is the drawdown time. ASCE (2001) recommends a target drawdown time of 40 hours.

In recent years, the use of sand filters has been questioned due to the high cost of operation. Even with pretreatment facilities, sediments can still quickly accumulate in the filter and must be removed to maintain filtration capacity. In addition, since the transport capacity of the downstream channel will generally exceed the sediment content of the effluent, the use of sand filters may actually increase downstream sediment degradation (i.e., bed erosion) (Wurbs and James, 2002). This latter problem may be experienced with other BMPs, such as detention and retention basins, as well.

It is also worth mentioning, that filtration devices such as sand filters have a major tradeoff; they can be highly effective, but they cannot accommodate large flow quantities. If used alone, flow must pass through the filter to be effectively treated. They are, therefore, limited to small- to medium-sized drainage areas and relatively low flows. Larger flows, caused by high intensity storms, will tend to bypass the filter. In some cases, simultaneous application of other BMPS (e.g., detention facility), can remedy this situation. For flow that is passed through the filter, a layer of organic matter, or biomass, will develop within the BMP, and the adsorption potential and removal efficiency of pollutants will actually improve over time as this biomass increases. This is especially true in areas of warmer climate.

Bioretention Basins and Constructed Wetlands: These BMPs rely primarily on biological processes for pollutant control. On average, they are highly effective for the removal of soluble pollutants (e.g., nutrients), suspended solids, and metals, but generally provide only limited water quantity control. In operation, the combination of a shallow, depressed pool and rich, emergent vegetation allow a natural food chain to develop. Thus, pollutant removal can occur through biological plant uptake, microbial decomposition, volatilization, filtering, settling, and other processes (Brown, et al., 1996). If maintained properly, they can promote wildlife habitat and are generally more aesthetically pleasing than other BMPs.

Wetlands and bioretention basins are similar in many respects; however, wetlands require more vegetation, may incur slightly larger depths, and are designed as a small-scale network of permanent micro-pools, channels, and marsh areas (Novotny, 2003). Thus, for urban areas, bioretention and bioinfiltration basins (see Figure 8-2) are generally more suitable than wetlands as they can be placed in parking lot islands and around buildings. Basin dimensions should be at least 15 ft (4.6 m) wide by 40 ft (12.2 m) long with side slopes of no steeper than 4H:1V. The minimum surface area required can be computed using Equation 8-4 if Z_b is the depth of planting soil and h_{max} is the maximum depth of water in the basin. The size of drainage area treated by one bioretention cell should not exceed one acre (0.4 ha). In addition, ponding depths should be no greater than 6 inches (15 cm) above the filter bed.

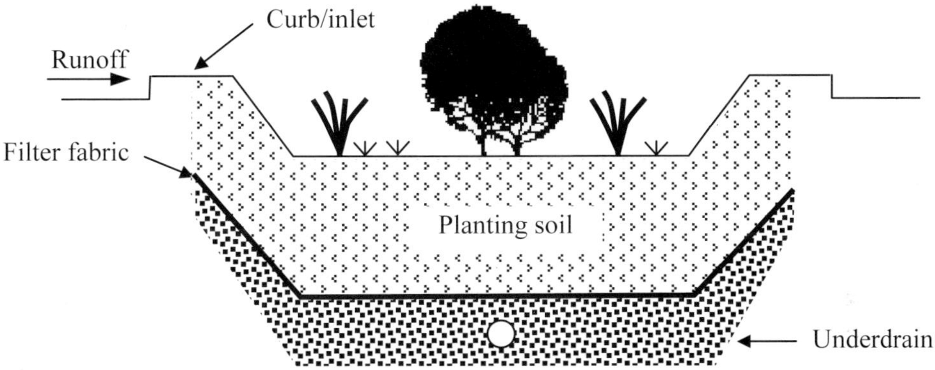

Figure 8-2: Typical bioretention cell (adapted from USEPA, 1999a)

Bioretention basins are not appropriate at locations where the water table is within 6 ft (1.8 m) of the ground surface or where surface slopes exceed approximately 20 percent. Significant attention must be paid to careful selection of plants and conditioning of special planting soils for development of adequate vegetation and promotion of infiltration (USEPA, 1999b). Generally, planting soils should extend to a depth between 1 and 4 ft (0.3 and 1.2 m) below the filter surface and should be covered with a layer of mulch. In many cases, an underdrain is used to release the retention volume being infiltrated. Note that both bioretention basins and wetlands have moderately high maintenance (e.g., watering, plant upkeep, and litter removal) requirements. They may also require pretreatment (e.g., filtering) facilities to remove coarse sediments that would degrade system performance. Currently, bioretention designs vary among applications, but as more data is collected over time, better and more cost-effective designs and standards should emerge.

Grass Swales: Swales are broad, shallow channels with a dense stand of vegetation covering the side slopes and bottom of the channel. They are wider than typical channels in order to accommodate design flows at depths below the height of vegetation. Grassed swales can effectively trap particulates for drainage areas up to approximately 10 ac (4 ha), but depending on the site conditions, may have a limited ability to reduce finer or soluble pollutants and control peak runoff rates through infiltration and reduction of flow velocities. As such, they are often used as pretreatment for other BMPs, and they may have filter media and an underdrain system added to further promote infiltration and treatment. Swales tend to be particularly common in residential areas and highway medians, where standing water is not desired. They are not recommended, however, in cases where peak runoff velocities exceed 3 fps (0.9 m/s). They should generally have longitudinal slopes as close to zero as the drainage design will permit, although two to six percent slopes are common. Lengths and widths should be at least 100 ft (30 m) and 2 ft (0.6 m), respectively, and swales should have residence times greater than five minutes. In addition, side slopes should be 3H:1V or flatter, and underlying soils should have a high infiltration capacity (Shueler, 1987; ASCE, 2001).

Standard open channel flow equations can be used in designing swales. However, the Manning roughness coefficient should be expressed as a function of the hydraulic radius, R, of the channel as follows (Chang and Cotton, 1988)

$$n = R^{1/6}\left[a_c + 19.97\log\left(R^{1.4}S^{0.4}\right)\right]^{-1} \tag{8-5}$$

where a_c is 15.8, 23.0, 30.2, 34.6, and 37.7 for vegetative retardance classes A, B, C, D, and E, respectively (see Table 8-1); and R is in ft.

Alternatively, using a mass balance of inflows and outflows, the length of swale, L, required for full infiltration of the design storm, can be expressed as

$$L = \frac{Q}{Pi_c} \tag{8-6}$$

where Q is the average flow rate, P is the wetted perimeter, and i_c is the final, saturated infiltration rate. Thus, for any cross-sectional shape, P can be related to Q via the Manning equation. For example, for trapezoidal swales, Wanielista et al. (1997) used the concept of the most efficient cross section to show that

$$L = \frac{K_u Q}{i_c}\left[B + 2\left\{\frac{K_n nQ\left(1+Z^2\right)^{1/3}}{S^{1/2}Z^{2/3}\,2\left[\left(1+Z^2\right)^{1/2} - Z\right]}\right\}^{3/8}\left(1+Z^2\right)^{1/2}\right]^{-1} \tag{8-7}$$

where K_u and K_n are constants equal to 43,200 and 1.068, respectively, in U.S. customary units and 360,000 and 1.0 in S.I. units; Q is in cfs (m^3/s); B is the bottom width in ft (m); n is the Manning roughness coefficient, which can be taken as approximately 0.20 for routinely mowed swales and 0.24 for infrequently mowed swales (ASCE, 1998); Z is the horizontal component of the side slope (i.e., ZH:1V); and i_c is in units of in/hr (cm/hr). For triangular swales

$$L = \frac{K_t Q^{5/8} Z^{5/8} S^{3/16}}{n^{3/8} \left(1 + Z^2\right)^{5/8} i_c} \tag{8-8}$$

Table 8-1: Vegetative covers by retardance class

Class	Cover	Condition
A	Weeping lovegrass	Excellent stand, 30 in (76 cm)
	Yellow bluestem	Excellent stand, 36 in (91 cm)
B	Kudzu	Very dense, uncut growth
	Bermuda grass	Good stand, 12 in (30 cm)
	Native grass mixture	Good stand, unmowed
	Weeping lovegrass	Good stand 24 in (61 cm)
	Alfalfa	Good stand, uncut, 0.91 ft (0.28 m)
C	Crabgrass	Fair stand, uncut, 10-48 in (25-120 cm)
	Bermuda grass	Good stand, mowed, 6 in (15 cm)
	Grass-legume mix	Summer, uncut, 6-8 in (15-20 cm)
	Kentucy bluegrass	Good stand, 6-12 in (15—30 cm)
D	Bermuda grass	Good stand, cut to 2.5 in (6 cm)
	Buffalo grass	Good stand, uncut 3-6 in (8-15 cm)
	Grass-legume mix	Fall, spring, uncut, 4-5 in (10-13 cm)
E	Bermuda grass	Good stand, cut to 1.5 in (4 cm)
	Bermuda grass	Burned stubble

Source: Adapted from Chang and Cotton (1988)

where K_t assumes a value of 21,032 in U.S. customary units and 151,361 in S.I. units. In some cases, the length required will be excessive or greater than the site allows. To overcome such a dilemma, small check dams or berms may be installed at intervals throughout the channel; this creates a wet swale in order to enhance treatment processes (i.e., settling, adsorption, uptake by vegetation, etc.) and provide storage for peak flow attenuation. Check dams might consist of several gabions (i.e., pervious, rock-filled basket) with a weir overflow. Material used in construction of dams should generally have a median diameter of 1 to 3 in (25 to 75 mm), and spacing of dams should not exceed the horizontal distance from the toe of the upstream dam to the same elevation at the top of the downstream dam.

Infiltration Systems: Infiltrating a portion of the runoff through coarse media and underlying soils reduces peak surface discharges to streams and collection systems and simultaneously recharges groundwater. It also allows for filtering and potential biodegradation of soluble and particulate pollutants. Commonly used systems that rely heavily on infiltration include porous pavements, infiltration basins (see Chapter 6), and infiltration trenches. The latter, shown in Figure 8-3, can serve drainage basins as large as 10 ac (4 ha) and consist of an excavated, permeable-lined trench that has been backfilled with stone having a median diameter of 1.5 to 3 in (40 to 80 mm) and porosity of 30 to 40 percent (Akan and Houghtalen, 2003). Runoff should either drain from the trench into native soil or enter an underdrain within 24 hours (i.e., D_T) for less permeable soils or within up to 72 hours for more pervious soils. Trench depths often range from 3 to 8 ft (1.0 to 2.5 m) for water quality control purposes and can be larger if water quantity is a design factor (Schueler, 1987; ASCE 2001). The maximum allowable depth, d_{max}, can be computed as (Akan and Houghtalen, 2003)

$$d_{max} = \frac{i_c D_T}{\eta} \tag{8-9}$$

where i_c is the final infiltration capacity of surrounding and underlying soils and η is the porosity of stone fill. Simultaneous consideration, however, should be given to the guideline that the bottom of the trench should lie at least 4 ft (1.2 m) above bedrock and at least 2 ft (0.6 m) above the high groundwater level (ASCE, 2001). Once the depth is determined, the remaining dimensions of the trench should be estimated, noting that the trench should store the runoff volume and rain falling directly on the trench, minus the quantity of water infiltrated into surrounding native soil while the trench is filled. For a rectangular trench having a bottom area of $W \times L$, this criterion can be expressed as

$$\eta(LWd) = V_c + (LW)P - (LW)i_c F_T \tag{8-10}$$

where P is the rainfall depth; and F_T is the trench fill time. If the trench is designed for water quality purposes only (i.e., capture of smaller, more frequent storms), only the capture volume need be considered on the right-hand side of Equation 8-10.

Similar to sand filters, the effectiveness of infiltration practices can be particularly poor if percolation rates are low or if sediments have not been adequately removed prior to groundwater recharge; problems with clogging can often occur quickly (Wurbs and James, 2002; Field and Sullivan, 2003). To prevent such problems, a filter strip can be installed around the perimeter of the trench. In addition, if groundwater in the area is used as water supply, care should be taken so that infiltration does not increase negative health impacts due to the migration of contaminants.

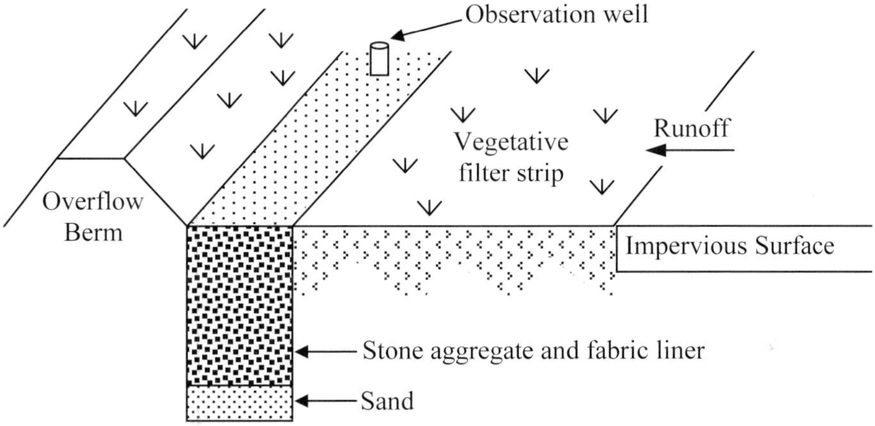

Figure 8-3: Typical infiltration trench (adapted from USEPA, 1999a)

Flow Control and Inlet Devices: These devices range from riprap to concrete flow spreaders, which dissipate kinetic energy at outfall locations, to water quality inlets and inserts (e.g., silt screens and oil/grit separators) that remove certain pollutants and floatable debris before they enter a collection system. Baffle boxes, installed in collection pipes for in-line removal of solids, are also considered within this category. While the devices are generally limited to serving smaller basins that are less than 5 ac (2 ha) in size, they are relatively simple to implement and can greatly improve stormwater control and pollutant removal efficiencies.

8.4.2 BMP Selection Factors

A variety of factors affect the screening and selection of BMPs to achieve identified goals. These commonly include:

- local regulations or requirements;
- desired pollution removal efficiency;
- political and public support;
- condition and designated use of receiving waters;
- existence of nearby substitute, unimpaired waters;

- drainage area size and corresponding land uses;
- climate, including average rainfall frequency, duration, and intensity;
- soil types;
- depth to groundwater;
- site topography;
- soil types and geologic character (e.g., bedrock and karst formations);
- availability of land;
- future development and land uses;
- availability of supplemental water to support vegetative BMPs;
- susceptibility to freezing;
- safety;
- ability of the BMP to treat multiple pollutants and problems; and
- periodic and long-term maintenance needs, including accessibility.

Two of the most influential factors that are not explicitly mentioned in the previous list are cost and benefits to be earned from selected BMPs. Ideally, selection of BMPs should be based on the relationship between these two factors, or a cost-benefit analysis. To aid in the analysis, the USEPA (1999a) has compiled information from the literature that provides estimates for capital and annual maintenance costs for some of the more common treatment BMPs (see Tables 8-2 and 8-3).

Unfortunately, while the data presented in these tables is useful in planning, it does not constitute enough information needed to make a comprehensive comparison of the cost effectiveness of different practices. The realized cost will likely include a variety of additional indirect and intangible costs. For example, it might consider resolution of local and regional social, economic, and political constraints, as well as a broad range of secondary environmental impacts.

Preparing a comparable list of specific benefits is similarly difficult due to the lack of clear and consistent definition of BMP performance and insufficient data; most data has focused on characterization of pollutant or discharge problems and not the quantifiable effectiveness of control measures. The USEPA has recently provided partial guidance with regard to assessing BMP effectiveness, and thus benefits. Table 8-4 summarizes reported pollutant removal efficiencies for some of the more common treatment BMPs (USEPA,

1999a). It should be noted, however, that attempts to broadly extend such data and related conclusions should be avoided since performance is heavily dependent on a host of site-specific factors such as the concentration of targeted pollutants, climate, seasonal temperature variation, and others.

Table 8-2: Typical base capital costs for BMPs

BMP	Cost [a] ($ per ft^3 of Treated Water)	Comments
Retention and Detention Basins	0.5 – 1.0	Cost range reflects economies of scale in designing BMP. The lowest unit cost represents approxmately150,000 ft^3 of storage, while the highest is approximately 150,000 ft^3. Dry detention basins are the least expensive design option.
Filter Strip	0.0 – 1.3	Based on cost per square foot, assuming 6 inches of storage in the filter. The lowest cost assumes that the buffer uses existing vegetation, and the highest cost assumes that sod was used.
Sand Filter	3.0 – 6.0	The range in costs is largely due to different possible designs. Surface and underground filters are among the most costly.
Bioretention Basin	5.3	This BMP is relatively constant in cost because it is usually designed as a constant fraction of the total drainage area.
Constructed Wetland	0.60 – 1.25	Costs are assumed to be approximately 25 percent more expensive than retention basins, primarily due to plant selection and sediment pretreatment requirements.
Grass Swale	0.50	Based on cost per square foot, assuming 6 inches of storage in the filter.
Infiltration Basin	1.3	Represents typical costs for a 0.25-acre infiltration basin.

[a] 1997 U.S. dollars; 1 $ft^3 = 2.83 \times 10^{-2}\ m^3$
Source: Adapted from USEPA (1999a)

For many pollution prevention and control practices, quantifiable measures of effectiveness have not been reported, as they are generally unattainable. Surrogate measures of effectiveness, such as the degree of change exhibited in people's activities or the degree of reduction in source pollutants are more typically used. These measures might include the number of public workshops held per year, periodic surveys of residents' recycling habits, or the volume of material recycled each month.

Table 8-3: Annual maintenance costs for BMPs

BMP	Annual Maintenance Cost (Percent of Construction Cost)
Detention Basin	< 1
Retention Basin	3 – 6
Sand Filter	11 – 13
Bioretention Basin	5 – 7
Constructed Wetland	2
Grass Swale	5 – 7
Infiltration Basin	1 - 10

Source: Adapted from USEPA (1999a)

Likewise, the effectiveness of water quantity controls is difficult to measure and broadly quantify due to the variability of watershed response to precipitation. The information that does exist, however, at least confirms a direct correlation between urbanization and degradation of receiving streams. Generally speaking, degradation is particularly accelerated when impervious area in the watershed exceeds approximately five to ten percent of the total area (Pitt and Clark, 2003). In some cases, it may be possible to investigate the local level of control necessary to compensate for impervious areas greater than these critical values.

8.4.3 The Role of Optimization

As focus has shifted toward the importance of a watershed-scale approach for mitigating stormwater impacts, recent work has shown that optimization and search methods can be effectively used in cost-effective design of certain treatment BMPs and in determining landscapes that are most conducive to pollutant reduction. The majority of this computational research is based on the coupling of various hydrologic and environmental simulation models with evolving search techniques. The search method iteratively, but strategically, determines decision variable values that maximize pollutant reduction goals or minimize cost required to meet water quality criteria. Here, decision variables might be land use allocation, BMP sizing and placement criteria, or other important, independent parameter. The simulation model is then called each time the search method requires information about the state (e.g., pollutant concentration in receiving streams) of the watershed in response to a prescribed set of decision variables.

Table 8-4: Estimated pollutant removal efficiency

	Typical Pollutant Removal in Percent				
	Suspended Solids	Nitrogen	Phosphorus	Pathogens	Metals
Detention Basins	30 -85	15 - 45	15 - 50	< 30	15 – 70
Retention Basins	50 - 98	30 - 98	30 - 98	< 30	50 – 98
Constructed Wetlands	50 - 80	< 30	15 - 45	< 30	50 – 80
Infiltration Basins	50 - 80	50 - 80	50 - 80	65 - 100	50 – 80
Grassed Swales	30 - 83	< 45	15 - 45	< 30	15 – 45
Filter Strips	50 - 80	50 - 80	50 - 80	< 30	30 – 65
Sand Filters	50 - 95	< 47	41 - 80	< 30	20 – 80

Source: Adapted from USEPA (1999a) and ASCE (2001)

The benefits of this integrative approach are discussed by Nicklow (2000) and include a reduced size and complexity of the overall optimization problem since the simulation model implicitly solves governing hydrologic and environmental constraints. At the same time, the complexity of the physical system is maintained and additional model simplifications are not typically necessary to reach an optimal or near-optimal result. Historically, the use of traditional, gradient-based optimization techniques within this overall approach was problematic; such techniques typically required significant derivative information about the simulation equations that was difficult to obtain efficiently, if at all. However, when used with advanced heuristic algorithms (i.e., genetic algorithms, evolutionary algorithms, and simulated annealing) that require no gradient information, the integrative approach is relatively efficient and can lead to very effective, basin-wide stormwater control.

Major research contributions in this field include that of Dorn et al. (1995), Harrell and Ranjithan (2003), Perez-Pedini et al. (2004), Zhen et al. (2004) and several others who have demonstrated a variety of methods that can be used to evaluate the most cost-effective, system-wide configuration of detention basins or similar BMPs to meet watershed pollutant removal levels. Likewise, researchers have applied multiobjective search methods to directly generate solutions for minimizing system cost while maximizing the effects of detention systems (Yeh and Labadie, 1997; Dorn and Ranjithan, 2003).

From a land-use perspective, Nicklow and Muleta (2001) developed a computational model to identify landscapes (i.e., land uses) and management practices, in both a spatial and temporal sense, that could be implemented to minimize erosion of rural lands. The method was subsequently expanded to yield multiple tradeoff solutions for the multiobjective case of maximizing landowner profit and minimizing non-point source pollution (Muleta and Nicklow, 2002; Bekele and Nicklow, 2004). With anticipated improvements in computer speed and capacity, hydrologic and environmental simulation models, and optimization and search methods, it is expected that the use of optimization in BMP design and overall watershed management will become more common and beneficial over time.

8.5 IMPLEMENTATION, MONITORING, AND ASSESSMENT

After choosing control practices, the watershed plan should be implemented in a phased approach. This will require a number of new elements that were not necessarily part of the planning phase. First, building upon the framework presented by Heathcote (1998), a single lead agency or entity should assume the role as advocate and facilitator for plan implementation with both the community and political representatives. There must also be a clear delineation of the roles and responsibilities of other stakeholders. In many cases, the successful implementation of BMPs and associated watershed management plan will hinge upon the understanding and the active participation of all stakeholders. Second, an estimate of time and costs associated with each project task should be outlined, monitored, and updated as needed to meet schedule and budgetary constraints. Finally, it is important to recognize relationships to existing regulatory or private planning processes (e.g., land use ordinances and flow monitoring programs) in order to minimize duplication of efforts and maximize the use of available information.

Once implemented, the management plan should be a dynamic document that is revised every three to six years to reflect the cyclic evaluation of the problem, goals, and control measures to be implemented. This reevaluation process is essentially part of an adaptive management framework whereby stakeholders acknowledge the inevitable uncertainty that existed in the planning phase with regard to data collection, hydrologic modeling, and the design of controls, as well as the variability of nature itself. Updated plans should consider recent changes in sewer operation, land use, population, and other factors that may affect future directions of planning efforts.

A critical aspect of updating a watershed management plan is monitoring and assessment to determine whether the initial problem is being resolved. Additionally, monitoring will reveal whether the solution to one problem unmasks another problem and whether initial goals were realistic. These assessment results should be communicated to stakeholders in order to sustain their interest and enthusiasm for the ongoing project.

8.6 SOLVED PROBLEMS

Problem 8.1 Problem identification

Identify five major categories of pollutants that are likely to be found in urban stormwater, along with their probable sources and associated impacts.

Solution

- Sediment and similar suspended solids: Sources include eroded land, weathering and wear of pavement, buildings, and other structures, construction practices, leaves and grass clippings, car washing, and atmospheric deposition. Solids produced from these activities adds to turbidity in water, thereby reducing available light for plant growth and affecting organisms that feed on those plants. Downstream deposition of solids can change morphology of receiving waters and severely limit growth of underlying organisms.
- Nutrients (phosphorus and nitrogen): Sources include organic matter, fertilizers, SSOs, CSOs, animal fecal matter, car wash detergents, and atmospheric deposition. Nutrients can stimulate excessive aquatic plant growth (e.g., weeds and algae) that may lead to significant fluctuations in dissolved oxygen concentrations.
- Heavy metals: Sources include SSOs, CSOs, vehicle wear and tear, atmospheric deposition, illegal discharges or spills, and weathering of buildings and other structures. Metals act as poisons to living organisms and tend to persist in the environment for a long period.
- Litter and debris: There are numerous sources that include pedestrians, vehicles, waste collection operators, leaves, and grass clippings. The most apparent impact is floating debris, which is aesthetically displeasing. Animals can be sickened or choke on the material.
- Microorganisms: Sources include fecal matter, SSOs, CSOs, and decaying organic matter. These organisms can contain significant bacteria and viruses, thus increasing the potential for waterborne illness.

Problem 8.2 Channel protection

Determine the appropriate size of riprap to be used in an alluvial channel having a slope of 2.1 percent. The maximum water depth is estimated as 3.0 ft.

Solution

The median diameter of material can be evaluated using Equation 8-1, as follows:

$$D_{50} = \frac{(62.4)(3.0)(0.021)}{5} = \underline{\underline{0.79\ ft}}$$

The riprap should further be graded from 0.2 × D_{50} to 2.0 × D_{50}, or 0.16 ft to 1.57 ft.

Problem 8.3 Filter strip

Determine the dimensions of a filter strip to capture a design flow of 0.5 cfs at constant rainfall intensity of 0.17 in/hr. The filter will have longitudinal slope of 0.045 ft/ft and Manning roughness coefficient of 0.20 (i.e., combination of very dense grass and shrubs).

Solution

Assuming a maximum depth of 1.0 inch, the unit-width flow is evaluated using Equation 8-3, as follows:

$$q = \frac{1.49}{0.20}\left(\frac{1.0}{12}\right)^{5/3} 0.045^{1/2} = 0.025\ cfs/ft$$

The minimum filter width, W, is then

$$W = \frac{Q}{q} = \frac{0.5}{0.025} = 20\ ft$$

Using a conservative ten-minute retention time, the corresponding length is computed as

$$L_f = \frac{(10)^{5/3}(0.17)^{2/3}(0.045)^{1/2}}{(0.899)(0.20)} = 16.8\ ft$$

which is less than the minimum suggested length of 20 ft. Therefore, the filter should be <u>at least 20 ft × 20 ft</u>, and, if land is available, an even larger length (i.e., 20 ft wide × 50 ft) could be specified to improve treatment performance.

Comments: Note that in most applications, intensity will vary according to time of concentration, or hydraulic retention time. Thus, unlike this example, the solution for filter strip length is more likely to require use of a site-specific intensity-duration-frequency relationship.

Problem 8.4 Sand filter

Design a sand filter to treat the water quality volume associated with a 4-ha, 75 percent impervious, commercial site. A pretreatment sedimentation basin is to be used for releasing runoff to the filter over a 24-hr period. Assume a drawdown time of 1.5 days and a coefficient of permeability of 1.1 m/d.

Solution

Assuming a 0.5 in (0.013 m) runoff for the water quality volume, the capture volume, V_c, is computed as

$$V_c = (0.013)(0.75)(40{,}000) = 390\ m^3$$

The sedimentation basin then releases flow at a rate of 390/24, or 16.25 m^3/hr, and flow will infiltrate through the sand at 390/36, or 10.83 m^3/hr. The daily volume of runoff that will accumulate on the sand surface is equal to

$$(16.25 - 10.83)(24) = 130\ m^3$$

at a maximum depth of $130/A_b$. Assuming that the average depth is half of the maximum depth and that the bed is 0.5 m deep, the minimum surface area of the bed can be computed as

$$A_b = \frac{(390)(0.5)}{1.1\left(\dfrac{130/A_b}{2} + 0.5\right)1.5} = 106.4\ m^2$$

Thus, a <u>1.5 m wide × 75 m long × 0.5 m deep filter</u> would satisfy the requirements of the site.

Problem 8.5 Bioretention basin

Determine the size of bioretention basin necessary to capture and treat 0.05 ac-ft (i.e., water quality volume) of runoff. The planting soil to be used has an infiltration capacity of 1.0 in/hr and extends to a depth of 4 ft below the filter bed.

Solution

Assuming a maximum ponding depth of 0.5 ft and that drawdown should occur within 40 hrs, the minimum area of the bed (i.e., invert) can be computed as

$$A_b = \frac{(0.05 \times 43{,}560)(4)}{(1/12)(0.5+4)(40)} = 580\ ft^2$$

So a 15 ft × 40 ft (i.e., 600 ft^2) basin would seem adequate. However, the temporary basin volume above the mulch layer should be checked to ensure that the water quality volume (i.e., 0.05 × 43560 = 2,178 ft^3) can be accommodated at reasonable depth. For a trapezoidal basin having a length *L*, width *W*, and side slopes of ZH:1V, the following expression can be solved for depth:

$$V = 2{,}178 = LWD + (L+W)ZD^2 + \frac{4}{3}Z^2D^3$$

Using this expression with $W = 15$ ft, $L = 40$ ft, and $Z = 8$ yields a depth of 1.5 ft which exceeds the maximum ponding limit of 0.5 ft, so the basin dimensions must be increased. For example, a basin that is 45 ft × 85 ft, with 8H:1V side slopes will store the capture volume at a satisfactory depth of 0.5 ft.

Problem 8.6 Grass swale

Design a grass swale to accommodate a design runoff of 0.5 m^3/s. Mowed Bermuda grass (retardance class D) is to be used and the depth should be no more than 0.3 m. Based on topography, the swale's longitudinal slope is to be 2.0 percent.

Solution

Using a trapezoidal section with side slopes of 5H:1V, the area and hydraulic radius for a depth of 0.3 m can be expressed as

$$A = (b+zy)y = [b+(5)(0.3)]0.3 = 0.3b + 0.45$$

and

$$R = \frac{(b+zy)y}{b+2y\sqrt{1+z^2}} = \frac{0.3b+0.45}{b+(2)(0.3)\sqrt{1+5^2}} = \frac{0.3b+0.45}{b+3.06}$$

Substituting these parameters into the Manning equation yields

$$Q = 0.5 = \frac{1.0}{n}\frac{(0.3b+0.45)^{5/3}}{(b+3.06)^{2/3}}0.02^{1/2}$$

This relationship can be solved for *b* according to assumed values of *n*. The value of *n*, however, must be consistent with Equation 8-5, written for class D retardance. For example, assuming a roughness coefficient of 0.04, the channel bottom width is computed as 0.14 m using the Manning equation. From the previous relationship for hydraulic radius, R is evaluated as 0.154 m, or 0.505 ft. An updated roughness coefficient can be computed using

$$n = 0.505^{1/6}\left[34.6 + 19.97 \log\left\{\left(0.505^{1.4}\right)\left(0.02^{0.4}\right)\right\}\right]^{-1} = 0.070$$

Thus a second iteration is necessary in which n = 0.070 is used to solve for b. The following summarizes the iterative solution, which yields a final bottom width of 0.83 m.

Assumed n	b (m)	R (m)	R (ft)	Computed n
0.04	0.14	0.154	0.505	0.07
0.07	1.03	0.186	0.61	0.061
0.061	0.77	0.178	0.584	0.063
0.063	**0.83**	0.18	0.591	0.063 (OK)

Velocity can then be checked by

$$\frac{Q}{A} = \frac{0.5}{\left[0.83 + (5)(0.3)\right]0.3} = 0.72\frac{m}{s}$$

which is in an acceptable range. Using a minimum retention time of 5 minutes, the length of the swale should be at least

$$L = (0.72)(5 \times 60) = 216 \ m$$

In summary, the swale is designed to have a trapezoidal cross section, should be at least 216 m long and have a bottom width of 0.83 m and side slopes of 5H:1V.

Problem 8.7 Grass swale

Determine the minimum length of trapezoidal grassed swale necessary to drain a section of highway. The swale should fully percolate the design flow of 1.0 cfs, which occurs over a period of 30 minutes. Assume a routinely mowed surface having a slope of 0.02 ft/ft and a saturated infiltration rate of 4.0 in/hr.

Solution

Assuming a trapezoidal swale having a bottom width of 1.0 ft and side slopes of 5H:1V, the required length can be computed as

$$L=\frac{(43{,}200)(1.0)}{(4.0)}\left[1.0+2\left\{\frac{(1.068)(0.20)(1.0)(1+5^2)^{1/3}}{(0.02)^{1/2}(5)^{2/3}\,2\left[(1+5^2)^{1/2}-5\right]}\right\}^{3/8}(1+5^2)^{1/2}\right]^{-1}$$

which yields $\underline{\underline{L = 471\text{ ft}}}$.

Problem 8.8 Grass swale

Solve Problem 8.7 using a triangular swale.

Solution

Using a 8H:1V side slope, Equation 8-7 yields

$$L=\frac{(21{,}032)(1.0)^{5/8}(8)^{5/8}(0.02)^{3/16}}{(0.20)^{3/8}(1+8^2)^{5/8}(4.0)}=\underline{\underline{1{,}247\ ft}}$$

Problem 8.9 Check dam

Assume that the swale length computed in Problem 8.8 exceeds the physical limitations of the site. Specify the additional storage volume that must be created by a series of check dams if the length of the swale is to be no longer than 1,000 ft.

Solution

The additional storage volume required can be evaluated by first letting L assume a value of 1,000 ft and solving Equation 8-8 for Q as follows:

$$Q=\left[\frac{Ln^{3/8}(1+Z^2)^{5/8}\,i_c}{K_t Z^{5/8}S^{3/16}}\right]^{8/5}=\left[\frac{(1{,}000)(0.20)^{3/8}(1+8^2)^{5/8}(4.0)}{(21{,}032)(8)^{5/8}(0.02)^{3/16}}\right]^{8/5}=0.70\ cfs$$

The additional storage that should be created through use of the dams is

$$(1.0-0.7)(30)(60)=\underline{\underline{540\ ft^3}}$$

Site and swale geometry should be evaluated to ensure that this volume exists without causing overflow and to determine the number of check dams needed.

Problem 8.10 Infiltration trench

An infiltration trench will be used to control the quality of runoff from a parking lot that is 120-ft wide and 200-ft long. The porosity of backfill material is 0.40 and the surrounding soil has an infiltration rate of 1.2 in/hr. Assuming a desired drawdown time of 48 hrs and that the depth to groundwater is 15 ft, determine the required trench dimensions.

Solution

The water quality volume to be captured is evaluated as

$$V_c = \left(\frac{0.5}{12}\right)(120)(200) = 1{,}000\ ft^3$$

The maximum trench depth can be determined using Equation 8-9,

$$d_{max} = \frac{(1.2/12)(48)}{0.40} = 12\ ft$$

which is sufficiently above the groundwater table. Since only water quality concerns are considered, Equation 8-10 can be expressed as

$$\eta(LWd) = V_c$$

Then, assuming a width of 6 ft, the required length is computed using

$$0.40(L \times 6 \times 12) = 1{,}000$$

or L = 34.7 ft. Thus, a trench that is 6 ft × 12 ft × 35 ft should effectively treat the runoff. A filter strip should be considered for the perimeter of the trench, and trench aggregate should range from 1.5 to 3.0 inches in diameter.

Problem 8.11 BMP selection

Consider a highly urbanized drainage area of 5 ac (2 ha) that is 80% impervious. The site houses two gas stations with large, paved parking areas. The available space for constructing treatment BMPs is limited to two to three percent of the total basin, so subsurface BMPs would generally be desirable. Unfortunately, low-permeability soils in the area limit the use of infiltration systems. Evaluate and select BMP alternatives to reduce suspended solids, oil, and grease loadings to the existing storm sewer and eventually to a nearby watercourse.

Solution

Important considerations and limitations for the site include:

- Subsurface BMPs with a modular design should be utilized to the extent possible so that land overlying the BMP can be utilized for other purposes;

- Infiltration systems should not be used due to the low permeability of surrounding soils;
- The selected BMP should be capable of operating under a range of hydraulic heads since the existing storm drain will limit the amount of hydraulic drop (i.e., elevation change) that can be made available within the new BMP; and
- Oil and grease will attach to suspended solids and float on top of stormwater runoff.

The treatments BMPs most likely to meet these criteria are an underground sand filter –likely to achieve a 50 to 90 percent removal of suspended solids - and an inlet device such as an oil/grit separator. These measures must, however, be examined in light of funding that is available for construction, operation, and maintenance. Note that sand filters are among the most costly alternatives with respect to both capital and maintenance expenditures. If funding is insufficient, other BMPs might need to be considered, even if they provide a lower level of performance. In addition to these treatment BMPs, pollution prevention measures in the form of education and training of site personnel, sweeping of streets and parking areas, and on-site containment measures should be considered. These will limit the amount of larger solids and debris that enter the sewer, minimize the amount of fuel spilled, and improve response actions in the event of a spill.

Problem 8.12 BMP selection

Consider a 1,000-ft length of four-lane, metropolitan highway that drains to a major river system. Flooding in the river is not a significant concern, but phosphorus and suspended solids loadings have reached detrimental levels and BMPs are being considered. The area for control measures is limited to the median and shoulder, and groundwater levels are in close proximity to the surface. The site is relatively flat and is highly visible to passersby. Identify a suitable BMP or combination of BMPs to reduce solids and phosphorus yields.

Solution

Treatment BMPs that are not well suited include detention and retention basins, wetlands, filter strips, and infiltration systems. Those that could be considered are bioretention cells, grass swales, and sand filters. Although bioretention facilities could potentially achieve high levels of treatment, they may be difficult to size and place given physical limitations of the site. The combination of a median swale and sand filter, however, would fit well into a linear highway configuration and should yield an adequate level of treatment. This combination is also quite aesthetically acceptable, although costs for the controls would likely be moderate to high. These treatment BMPs should be implemented simultaneously with a litter control program (i.e., adopt-a-highway). Street sweeping on a large highway is probably not a feasible option due to safety considerations.

REFERENCES CITED

Akan, A.O., and R.J. Houghtalen, *Urban Hydrology, Hydraulics, and Stormwater Quality*, Wiley, Hoboken, NJ, 2003.

American Society of Civil Engineers (ASCE), *Urban Water Quality Management*, ASCE Manuals and Reports on Engineering Practice No. 87 and Water Pollut. Control Fed. Manual of Practice 23, New York, NY, 1998.

American Society of Civil Engineers (ASCE), *Guide for Best Management Practice Selection in Urban Developed Areas*, ASCE Urban Water Infrastructure Management Committee's Task Committee for Evaluating BMPs, Reston, VA, 2001.

American Society of Civil Engineers (ASCE), *National Stormwater BMP Database*, <http://www.asce.org/community/waterresources/nsbmpdb.cfm>, Reston, VA, 2004.

Bekele, E.G., and J.W. Nicklow, "Multi-Objective Evolutionary Algorithms for Cost-Effective Management of Ecosystem Services in a Watershed." *Proc. World Water and Environmental Resources Congress*, ASCE, Salt Lake City, UT, 2004.

Brown, S.A., S.M. Stein, and J.C. Warner, "Urban Drainage Design Manual," Hydraulic Engineering Circular No. 22, FHWA-SA-96-078, Federal Highway Administration, U.S. Department of Transportation, Washington, DC, 1996.

Chang, Y.H., and G.K. Cotton, "Design of Roadside Channels with Flexible Linings," Hydraulic Engineering Circular No. 15, FHWA-IP-87-7, Federal Highway Administration, U.S. Department of Transportation, Washington, DC, 1988.

Claytor, R., and T. Schueler, *Design of Stormwater Filtering Systems*. Prepared for the Chesapeake Research Consortium. Center for Watershed Protection. Ellicott City, MD, 1996.

Debo, T.N., and A.J. Reese, *Municipal Storm Water Management*, Lewis Publishers, CRC Press, Boca Raton, FL, 1995.

Dorn, J.L., S.R. Ranjithan, S. Liehr, and R.C. Borden, "Genetic Algorithm approach to water supply watershed management," *Proc. of 2nd Congress on Computing in Civil Engrg.*, ASCE, New York, NY, 971-978, 1995.

Dorn, J. L., and S. Ranjithan. "Evolutionary Multiobjective Optimization in Watershed Water Quality Management," in EMO 2003, LNCS 2632, ed. by C.M. Fonseca et al., Springer-Verlag Berlin, 2003.

Field, R., and D. Sullivan, "Management of Wet Weather Flow in the Urban Watershed," in *Wet Weather Flow in the Urban Watershed*, ed. by R. Field and D. Sullivan, Lewis Publishers, Boca Raton, FL, 2003.

Harrell, L.J., and S.R. Ranjithan, "Detention Pond Design and Land Use Planning for Watershed Management," *J. of Water Res. Plan. and Mgmt.*, ASCE, vol. 129, no. 2, 98-106, 2003.

Heathcote, I.W., *Integrated Watershed Management*, Wiley, New York, NY, 1998.

Muleta, M.K., and J.W. Nicklow, "Evolutionary Algorithms for Multiobjective Evaluation of Watershed Management Decisions." *J. of Hydroinformatics*, IWA, vol. 4, no. 2, 83-97, 2002.

Nicklow, J.W., "Discrete-Time Optimal Control for Water Resources Engineering and Management," *Water International*, IWRA, vol. 25, no. 1, 89-95, 2000.

Nicklow, J.W., and M.K. Muleta, "Watershed Management Technique to Control Sediment Yield in Agriculturally Dominated Areas," *Water International*, IWRA, vol. 26, no. 3, 435-443, 2001.

Normann, J.M., *Design of Stable Channels with Flexible Linings*, Federal Highway Administration Circular 15, U.S. Dept. of Transportation, Washington, DC, 1975.

Novotny, V., *Water Quality: Diffuse Pollution and Watershed Management*, Wiley, New York, NY, 2003.

Pearse, P.H., F. Bertrand, and J.W. MacLaren, *Currents of Change: Final Report of the Inquiry on Federal Water Policy*, Queen's Printer, Ottawa, 1985.

Perez-Pedini, C., R.M. Vogel, J. Limbrunner, and J. Durant, "Optimal Number and Location of BMPs for Stormwater Management," *Proc. of World Water and Environ. Res. Congress*, ASCE, Reston, VA, 2004.

Pitt, R., and S. Clark, "Emerging Stormwater Controls for Critical Source Areas," in *Wet Weather Flow in the Urban Watershed*, ed. by R. Field and D. Sullivan, Lewis Publishers, Boca Raton, FL, 2003.

Shueler, T.B., *Controlling Urban Runoff: A Practical Manual for Planning and Designing Urban BMPs*, Washington Metropolitan Water Resources Planning Board, Washington, DC, 1987.

Soil Conservation Service (SCS), "Urban Hydrology for Small Watersheds," Technical Release 55, U.S. Dept. of Agriculture, Washington, DC, 1986.

U.S. Environmental Protection Agency (USEPA), "Preliminary Data Summary of Urban Storm Water Best Management Practices," EPA 821-R-99-012, Washington, DC, 1999a.

U.S. Environmental Protection Agency (USEPA), "Storm Water Technology Fact Sheet: Bioretention," EPA 832-F-99-012, Washington, DC, 1999b.

Wanielista, M., R. Kersten, and R. Eaglin, *Hydrology: Water Quantity and Quality Control*, 2nd ed., Wiley, New York, NY, 1997.

Wurbs, R.A., and W.P. James, *Water Resources Engineering*, Prentice Hall, Upper Saddle River, NJ, 2002.

Yeh, C.-H., and J.W. Labadie, "Multiobjective Watershed-Level Planning of Stormwater Detention Systems," *J. of Water Res. Plan. and Mgmt.,* ASCE, vol. 123, no. 6, 336-343, 1997.

Zhen, X.-Y., S.L. Yu, and Lin, J.-Y., "Optimal Location and Sizing of Stormwater Basins at Watershed Scale, *J. of Water Res. Plan. and Mgmt.,* ASCE, vol. 130, no. 4, 339-347, 2004.

APPENDIX A
Nomenclature

Variable	Definition	Dimensions
A	Area (Flow, Drainage)	L^2
A	Alkalinity of cement-bonded material (Chapter 5)	[ ]
A_b	Minimum basin surface area (Chapter 6)	L^2
$A_{c,t}$	Cumulative watershed area contributing at time t	L^2
A_o	Orifice Area (Chapter 6)	L^2
A_T	Total watershed area	L^2
B	Top width (Chapter 3)	L
b	Bottom width of a channel (chapter 3)	L
BOD	Biochemical oxygen demand	M/L^3
BOD_5	Five day BOD	M/L^3
BOD_L	Ultimate BOD	M/L^3
C	Runoff coefficient	[ ]
C	Chezy coefficient	[ ]
c	Wave celerity (Chapter 4)	L/T
C	Constituent concentration	M/L^3
C	USLE land cover and management factor	[ ]
C	Average rate of corrosion penetration (Chapter 5)	L/T
C	Weir coefficient (Chapter 6)	[ ]
C_1, C_2, C_3	Muskingum coefficients (Chapter 4)	[ ]
C_c	Composite runoff coefficient	[ ]
C_e	Constituent concentration at junction exit	M/L^3
C_H	Hazen-Williams coefficient	[ ]
c_k	Kinematic wave celerity	L/T
CN	Curve number	[ ]
C_n	Courant number	[ ]
C_o	Orifice coefficient (Chapter 6)	[ ]
C_P	Unit hydrograph peaking coefficient	[ ]
C_t	Watershed storage coefficient	[ ]
C_v	Volumetric sediment concentration	L^3/L^3
D	Diameter of a circular channel	L
D	Hydraulic depth	L
d	Maximum stable depth of flow (Chapter 8)	L
D_{50}	Median diameter of a riprap (Chapter 8)	L
D_{gr}	Grain size	[ ]
d_{max}	Maximum allowable trench depth	L

Variable	Definition	Dimensions
DS	Dissolved sulfide concentration	M/L^3
D_T	Drawdown time	T
E	Weight of solids contained in runoff	M
E_0	Initial weight of solids	M
EBOD	Effective BOD	M/L^3
E_{min}	Minimum specific energy	L
F	Flow force	F
f	Darcy-Weisbach friction factor	[ ]
f_b	Friction factor for sediment bed	[ ]
f_c	Composite Dracy-Weisbach friction factor	[ ]
f_g	Darcy-Weisbach friction factor corresponding to a grain shear stress	[ ]
F_g	Grain mobility factor	[ ]
F_r	Froude Number	[ ]
F_s	Effective sediment mobility	[ ]
F_T	Trench fill time	T
f_w	Clean-pipe friction factor	[ ]
g	Gravitational acceleration	L/T^2
H	Total head on the weir (Chapter 6)	[ ]
h	Stage (Chapter 6)	L
h	Allowable depth (Chapter 8)	L
h_L	Head loss	L
h_{max}	Maximum depth of water over the filter surface	L
h_p	Head supplied by pump	L
H_u , H_d	Total energy heads at entrance and exit of a pipe, respectively	L
i	Rainfall intensity	L/T
i_{30}	Maximum average 30-minute rainfall intensity	L/T
I_a	Initial abstraction	L
i_c	Final soil infiltration rate	L/T
I_p	Peak discharge of inflow hydrograph (Chapter 6)	L^3/T
I_t	Inflow at time t	L^3/T
j	Proportion of dissolved sulfide present as H_2S	[ ]
K	USLE soil erodibility factor	M/L^2
k	Proportionality constant (Chapter 5)	L^{-1}
k	Coefficient of efficiency for acid reaction	[ ]
k_1	Rate constant	T^{-1}
K_p	Infiltration capacity of a sand filter	L/T
k_s	Equivalent sand grain roughness	L
L	Length	L
L_f	Minimum width of filter strip	L

Variable	Definition	Dimensions
L_s	Length-Slope factor	[ ]
M	Mass	M
M	Specific momentum (Chapter 3)	L^3
m	Sulfide loss coefficient (Chapter 5)	[ ]
M	Sulfide flux coefficient for force mains (Chapter 5)	T/L
m	Outflow adjustment factor (Chapter 6)	[ ]
M'	Effective sulfide flux coefficient for gravity pipes (Chapter 5)	L/T
n	Manning's coefficient	[ ]
N	Yen and Chow's overland texture factor	[ ]
N_s	Number of sources being traced (Chapter 5)	[ ]
O_p	Peak flow of the outflow hydrograph (Chapter 6)	L^3/T
P	Population	[ ]
P	Total rainfall depth (Chapter 2)	L
P	Wetted Perimeter (Chapter 3)	L
p	Flow pressure	F/L^2
P_b	Wetted perimeter of sediment bed	L
P_e	Depth of rainfall excess	L
P_E	Erosion control practice factor	[ ]
P_w	Pipe wall wetted perimeter	L
Q	Discharge (flow)	L^3/T
q	Unit width flow	L^3/T
Q_{base}	Accumulated peakable sanitary load	L^3/T
Q_n	Direct runoff hydrograph ordinate	L^3/T
Q_p, Q_{peaked}	Peak runoff and peak sanitary load, respectively	L^3/T
R	Constant linear reservoir parameter (Chapter 2)	[ ]
R	Hydraulic radius (Chapter 3)	L
r	Total rainfall depth (Chapter 5)	L
Re	Reynolds number	[ ]
R_r	Rainfall factor	[ ]
S	Specific gravity	[ ]
S	Slope	L/L
S	Potential maximum retention	L
S	Specific gravity of sediment particles	[ ]
S	Total sulfide concentration	M/L^3
S	Basin storage (Chapter 6)	L^3
S_0	Channel bed slope	L/L
S_f	Friction slope	L/L

Variable	Definition	Dimensions
S_{lim}	Limiting sulfide concentration	M/L^3
T	Flow age (Chapter 5)	T
T	Wastewater temperature	
t	Flow travel time	T
t_b	Base time of hydrograph (Chapter 6)	T
t_c	Time of concentration	T
T_d	Drain time (Chapter 6)	T
t_L	Lag time	T
t_{La}	Adjusted lag time	T
t_p	Time to peak	T
t_r	Duration of excess rainfall	T
T_r	Time associated with the rising limb of a hydrograph	T
t_{ra}	Alternative unit hydrograph duration	T
t_t	Time of concentration through filter strip	T
U	Unit hydrograph ordinate	L^3/T
V	Volume	L^3
V	Flow velocity	L/T
V_c	Capture volume (Chapter 8)	L^3
V_L	Limiting flow velocity	L/T
V_r	Total volume of runoff (Chapter 6)	L^3
V_s	Detention storage volume (chapter 6)	L^3
V_T	Velocity at the threshold of bed load movement	L/T
W	Weight	F
w	Particle settling velocity	L/T
W_{50}, W_{75}	Unit hydrograph widths at discharge equal to 50 and 75% of the peak discharge, respectively	T
W_b	Width of the sediment bed	L
W_e	Effective width of the sediment bed	L
y	Flow depth	L
y_c	Critical depth	L
y_n	Normal depth	L
y_s	Mean depth of sediment over the pipe invert	L
z	Elevation head	L
Z_b	Depth of the sand bed	L
γ (gamma)	Unit weight of fluid	$F/L^3 = M/(L^2T^2)$
η (eta)	Transport parameter for a continuous bed	[ ]
θ_t (theta)	Transition coefficient for the particle Reynolds number	[ ]

Variable	Definition	Dimensions
θ_w (theta)	Specific retention/ initial water content of the soil	[ ]
μ (mu)	Dynamic viscosity	M/(L-T)
ν (nu)	Kinematic viscosity	L^2/T
ρ (rho)	Fluid density	M/L^3
τ (tau)	Shear stress	F/L^2
Φ (Phi)	Soil porosity (Chapter 6)	[ ]
Φ_{sw} (phi)	Flux of hydrogen sulfide to the pipe wall	$M/(L^2\text{-}T)$

APPENDIX B
Unit Conversions

Both SI and English units are used throughout this text. Tables B-1 to 7 give conversion factors for common units for different quantities.

Table B-1: Unit conversion tables for common units (length).

	Equivalent					
Unit	**Millimeter**	**Inch**	**Foot**	**Meter**	**Kilometer**	**Mile**
Millimeter	1	0.03937	0.003281	0.001	1E–6	0.6214E–6
Inch	25.4	1	0.0833	0.0254	25.40E –6	15.78E-6
Foot	304.8	12	1	0.3048	304.8E-6	189.4E-6
Meter	1 000	39.37	3.281	1	0.001	621.4E-6
Kilometer	1 000 000	39 370	3 281	1 000	1	0.6214
Mile	1 609 000	63 360	5 280	1 609	1.609	1

* Multiply the amount of the given unit by the equivalent value (per unit amount of given unit) to obtain the equivalent amount of the equivalent unit (e.g., 1.50 m x 39.37 inch/m = 59.1 inch)

Table B-2: Unit conversion tables for common units (volumetric discharge, volume/time).

	Equivalent					
Unit	**Gallon/min**	**Liter/sec**	**Acre-ft/day**	**Cu. foot/sec**	**Million gal/day**	**Cu. meter/sec**
Gallon/min	1	0.06309	0.004419	0.002228	0.001440	63.09E-6
Liter/sec	15.85	1	0.07005	0.03531	0.02282	0.001
Acre-ft/day	226.3	14.28	1	0.5042	0.3259	0.01428
Cu. foot/sec	448.8	28.32	1.983	1	0.6463	0.02832
Million gal/day	694.4	43.81	3.069	1.547	1	0.04381
Cu. meter/sec	15 850	1 000	70.04	35.31	22.82	1

Table B-3: Unit conversion tables for common units (flow area).

Unit	Equivalent		
	Sq. inch	**Sq. foot**	**Sq. meter**
Sq. inch	1	0.006944	645.2E-6
Sq. foot	144	1	0.09290
Sq. meter	1 550	10.76	1

Table B-4: Unit conversion tables for common units (drainage area).

Unit	Equivalent				
	Sq. meter	**Acre**	**Hectare**	**Sq. kilometer**	**Sq. mile**
Sq. meter	1	0.0002471	0.0001	0.000001	0.0000003861
Acre	4 046.94	1	0.4046856	0.0040469	0.0015625
Hectare	10 000	2.4710538	1	0.01	0.003861
Sq. kilometer	1 000 000	247.11	100	1	0.3861022
Sq. mile	2 589 988.11	640	259.0	2.59	1

Table B-5: Unit conversion tables for common units (volume).

Unit	Equivalent					
	Cu. inch	**Liter**	**U.S. gallon**	**Cu. foot**	**Cu. meter**	**Acre-foot**
Cu. inch	1	0.01639	0.004329	5.78.7E-6	16.39E-6	13.29E-9
Liter	61.02	1	0.2642	0.03531	0.001	810.6E-9
U.S. gallon	231.0	3.785	1	0.1337	0.003785	3.068E-6
Cu. foot	1 728	28.32	7.481	1	0.02832	22.96E-6
Cu. meter	61 020	1 000	264.2	35.31	1	810.6E-6
Acre-foot	75.27E6	1 233 000	325 900	43 560	1 233	1

Table B-6: Miscellaneous conversions.

Dimension	SI unit	Multiplier	Equivalent English unit
Velocity	Meter/sec	3.281	Ft/s
Mass	Kilogram (kg)	0.06852	Slugs
Force	Newton (N)	0.2248	Pound force
Pressure	Pascal (Pa)	0.0001450	Pound force/sq. inch
Pressure	Pascal (Pa)	0.02089	Pound force/sq. foot
Power	Watt (W)	0.001341	Horsepower = ft lb/s
Energy	Joule (J)	0.73746	Ft - lb

Table B-7: Other convenient conversion factors.

Temperature

$T(^oC) = 5 \times [T(^oF) - 32] / 9$

$T(^oF) = 32 + 9 \times [T(^oC)] / 5$

$T(^oK) = T(^oC) + 273$

$T(^oK) = 5 \times [T(^oF) + 460] / 9$

$T(^oK) = 5 \times T(^oR) / 9$

Specific Weight

$1\ lb/ft^3 = 157.1\ N/m^3$

^oR = degree Rankine
^oK = degree Kelvin
^oF = degree Fahrenheit
^oC = degree Celsius

INDEX